T0313157

Volume 3

The Innovation Tools Handbook

Creative Tools, Methods, and Techniques that Every Innovator Must Know

Volume 3

The Innovation Tools Handbook

Creative Tools, Methods, and Techniques that Every Innovator Must Know

Edited by **H. James Harrington** ● **Frank Voehl**

CRC Press
Taylor & Francis Group
Boca Raton London New York

CRC Press is an imprint of the
Taylor & Francis Group, an **informa** business

A PRODUCTIVITY PRESS BOOK

CRC Press
Taylor & Francis Group
6000 Broken Sound Parkway NW, Suite 300
Boca Raton, FL 33487-2742

First issued in hardback 2019

ISBN-13: 978-1-4987-6053-9 (hbk)

Library of Congress Cataloging-in-Publication Data

Names: Harrington, H. J. (H. James), editor. | Voehl, Frank, 1946- editor.
Title: The innovation tools handbook / H. James Harrington and Frank Voehl, editors.
Description: Boca Raton, FL : CRC Press, 2016- | Includes bibliographical references and index.
Identifiers: LCCN 2015042020 | ISBN 9781498760492 (vol. 1)
Subjects: LCSH: Technological innovations--Management. | Diffusion of innovations--Management. | New products.
Classification: LCC HD45 .I53795 2016 | DDC 658.4/063--dc23
LC record available at http://lccn.loc.gov/2015042020

When you have lived as long as I have, you have a right to look back and identify the things that are really important, which are the individuals who have helped make your life worth living and make it worthwhile getting up every morning. In my life, there have certainly been three people who have had a major impact upon the worthwhile things I've done and who have helped me get through the many mistakes I've made. These three people are my wife, Marguerite, my son, Jim, and my mother, Carrie. But these three people are all part of my family. Besides my immediate family, there have been a few other people who have overlooked my weaknesses and encouraged me to do more. People who continuously went out of their way to encourage me when I was down.

These people were not just fair-weather friends but are with me no matter what the weather is like. I dedicate this book to my all-weather friends—Candy Rogers, Chuck Mignosa, and Frank Voehl. Thanks for always being there to help me, sympathize with me, and encourage me.

Contents

Foreword

This book is the second of three volumes designed to provide its readers with the tools and methodologies that all innovators should be familiar with and able to use. These are the outputs from the Tools and Methodologies Working Group of the International Association of Innovative Professionals (IAOIP). The working group was made up of the following individuals:

- H. James Harrington, chairman
- Frank Voehl, co-chairman
- Yared Akalou
- Sifer Aseph
- Scott Benjamin
- Carl Carlson
- Gül Aslan Damcı
- Richard Day
- Lisa Friedman
- Thomas Gaskin
- Dallas Goodall
- Luis Guedes
- Paul Hefner
- Dana Landry
- Elena Litovinskaia
- Nikolaos Machairas
- Thomas Mazzone
- Chad McAllister
- Pratik Mehta
- Dimis Michaelides
- Howard Moskowitz
- Michael Phillips
- Jose Carlos Arce Rioboo
- Achmad Rundi
- Robert Sheesley
- Max Singh
- Nithinart Sinthudeacha
- Henryk Stawicki
- Maria Thompson
- Hongbin Wang
- David Wheeler
- Jay van Zyl

The mission statement for the Tools and Methodology Working Group is

Using the expertise and experience of the organization's members and literature research, the working group will define the tools and methodologies that are extensively used in support of the innovation process. The working group will narrow the comprehensive list of tools and methodologies to a list of the ones that are most frequently used in the innovative process and which are the ones that innovative professionals should be confident in using effectively. For each tool and methodology, the working group will prepare a write-up that includes its definition, when it should be used, how to use it, examples of how it has been used, and a list of 5 to 15 questions that can be used to determine if an individual understands the tool or methodology.

To accomplish this mission, the working group studied the literature that was currently available to define tools and methodologies that were presently proposed or being used. They also contacted numerous universities that were teaching classes on innovation or entrepreneurship to determine what tools and methodologies they were promoting. In addition, they contacted individual consultants who are providing advice and guidance to organizations in order to identify tools and methodologies they were recommending. As a result of this research, a list of more than 200 tools and methodologies was identified as being potential candidates for the innovative professional.

The group then sent surveys out to leading innovative lecturers, teachers, and consultants, asking them to classify each tool and/or methodology into one of the following categories:

- This tool or methodology is used on almost all the innovation projects = 4 points.
- This tool or methodology is used on a minimum of two out of five innovation projects = 1 point.
- This tool or methodology is seldom if ever used on innovative product projects = 0 point.
- Not familiar with the tool or methodology = –1 point.
- Never used or recommended this tool or methodology in doing innovation projects = –4 points.

We calculated the priority for each of the tools/methodologies by assigning a point value for each answer. The guidelines that we followed are

- Plus 4 points for a tool/methodology that was always used
- Plus 1 point for a tool/methodology that is being used at least two out of five projects
- No points for a tool/methodology that was seldom used
- Minus 1 for a tool/ methodology that the expert had never heard of
- Minus 4 points for a tool/methodology that the expert never used

Our goal was to define 50 of the most effective or most frequently used tools/methodologies by the innovative practitioner. We ended up with the 76 tools/methodologies that are the most effective or the most frequently used tools/methodologies by the innovative practitioner (professional).

We then submitted the selected 76 tools/methodologies to a group of 28 practicing innovators, asking them to write a chapter on one or more of the tools/methodologies.

When we assembled the 76 chapters, we ended up with a manuscript of about 1000 pages. After a discussion with the book's editors and key people in the Tools and Methodologies Working Group, it was decided to divide the book up into the following three books:

- Creative tools/methodologies that every innovator should master
- Evolutionary or improvement tools/methodologies that every innovator should master
- Organizational/operational tools/methodologies that every innovator should master

On the basis of these three breakdowns, we went out again to innovative experts asking them to classify each tool as falling into one of the three categories. We soon realized that many of the tools were used in more than one category, so we asked the experts to classify the category that the tool is primarily used in and indicate which categories the tool/methodology could also be used in. On the basis of this study, we divided the manuscript into three books:

- *Organizational and Operational Tools, Methods, and Techniques That Every Innovator Must Know*
- *Evolutionary and Improvement Tools That Every Innovator Must Know*
- *Creative Tools, Methods, and Techniques That Every Innovator Must Know*

Each book contains the tools/methodologies that were rated as primarily used in that category. The results of this study can be seen in Table F.1.

Although the vast majority of the patents and solutions fall into the evolutionary category, there are about 5% to 10% that are truly creative, unique, innovative, and original. Some organizations believe and base their business strategy on taking advantage of the market that is looking for something that is a little better rather than something that is different. There certainly is a lot less risk in coming out with a new product or service that is directed at an already established market rather than entering into an undefined market. These organizations that follow the already

TABLE F.1

Usage Classification for the Primary Innovative Tools and Methodologies

List of the Most Used and/or Most Effective Innovative Tools and Methodologies in Alphabetical order

Volume 1: Organizational and/or operational IT&M

Volume 2: Evolutionary and/or improvement IT&M

Volume 3: Creative IT&M

	IT&M	Volume 3	Volume 2	Volume 1
1.	5 Why questions	S	P	S
2.	76 Standard Solutions	P	S	
3.	Absence thinking	P		
4.	Affinity diagram	S	P	S
5.	Agile innovation	S		P
6.	Attribute listing	S	P	
7.	Benchmarking		S	P
8.	Biomimicry	P	S	
9.	Brain-writing 6–3–5	S	P	S
10.	Business case development		S	P
11.	Business plan	S	S	P
12.	Cause-and-effect diagrams		P	S
13.	Combination methods	P	S	
14.	Comparative analysis	S	S	P
15.	Competitive analysis	S	S	P
16.	Competitive shopping		S	P
17.	Concept tree (concept map)	P	S	
18.	Consumer co-creation	P		
19.	Contingency planning		S	P
20.	CO-STAR	S	S	P
21.	Costs analysis	S	S	P
22.	Creative problem solving model	S	P	
23.	Creative thinking	P	S	
24.	Design for tools		P	
	Subtotal Number of Points	7	7	10

(Continued)

TABLE F.1 (CONTINUED)

Usage Classification for the Primary Innovative Tools and Methodologies

List of the Most Used and/or Most Effective Innovative Tools and Methodologies in Alphabetical order

Volume 1: Organizational and/or operational IT&M

Volume 2: Evolutionary and/or improvement IT&M

Volume 3: Creative IT&M

	IT&M	Volume 3	Volume 2	Volume 1
25.	Directed/focused/structured innovation	P	S	
26.	Elevator speech	P	S	S
27.	Ethnography	P		
28.	Financial reporting	S	S	P
29.	Flowcharting		P	S
30.	Focus groups	S	S	P
31.	Force field analysis	S	P	
32.	Generic creativity tools	P	S	
33.	HU diagrams	P		
34.	I-TRIZ	P		
35.	Identifying and engaging stakeholders	S	S	P
36.	Imaginary brainstorming	P	S	S
37.	Innovation blueprint	P		S
38.	Innovation master plan	S	S	P
39.	Kano analysis	S	P	S
40.	Knowledge management systems	S	S	P
41.	Lead user analysis	P	S	
42.	Lotus blossom	P	S	
43.	Market research and surveys	S		P
44.	Matrix diagram	P	S	
45.	Mind mapping	P	S	S
46.	Nominal group technique	S	P	
47.	Online innovation platforms	P	S	S
48.	Open innovation	P	S	S
49.	Organizational change management	S	S	P
50.	Outcome-driven innovation	P		
	Subtotal Number of Points	15	4	7

(*Continued*)

TABLE F.1 (CONTINUED)

Usage Classification for the Primary Innovative Tools and Methodologies

List of the Most Used and/or Most Effective Innovative Tools and Methodologies in Alphabetical order

Volume 1: Organizational and/or operational IT&M

Volume 2: Evolutionary and/or improvement IT&M

Volume 3: Creative IT&M

	IT&M	Volume 3	Volume 2	Volume 1
51.	Plan–do–check–act	S	P	
52.	Potential investor present	S		P
53.	Proactive creativity	P	S	S
54.	Project management	S	S	P
55.	Proof of concepts	P	S	
56.	Quickscore creativity test	P		
57.	Reengineering/redesign		P	
58.	Reverse engineering	S	P	
59.	Robust design	S	P	
60.	S-curve model		S	P
61.	Safeguarding intellectual properties			P
62.	SCAMPER	S	P	
63.	Scenario analysis	P	S	
64.	Simulations	S	P	S
65.	Six Thinking Hats	S	P	S
66.	Social networks	S	P	
67.	Solution analysis diagrams	S	P	
68.	Statistical analysis	S	P	S
69.	Storyboarding	P	S	
70.	Synectics	S	S	P
71.	Systems thinking	P		
72.	Tree diagram	S	P	S
73.	TRIZ	P	S	
74.	Value analysis	S	P	S
75.	Value propositions	S		P
76.	Visioning	S	S	P
	Subtotal Number of Points	7	12	7

(P) Priority Rating	Creative	Evolutionary	Organizational
Total	29	23	24

IT&M in creativity book: 29

IT&M in evolutionary book: 23

IT&M in organizational book: 24

Note: IT&M, innovative tools and/or methodologies; P, primary usage; S, secondary usage; blank, not used or little used.

established trend play it safe; they have very few product failures but they also are not wildly successful. There are many economists who feel that the *follow the leader* concept is the best management strategy, and it probably is for those companies that are content to be good but will never be great. But if all organizations adapted this approach, there soon would be no one to follow and the world would be left with a bunch of humdrum products as organizations try to grab their share of a collapsing market. Work would be filled with monotony rather than excitement. These are the companies that are content to come out with the new model with a little additional chrome or a minor additional feature. Thank the Lord that there are a few companies that are willing to take a chance on the future and growing a new market for a new and exciting product/service. These are the organizations that are truly great. These are the ones that attract the best salespeople, the best engineers, and the most innovative managers. They make up that 10% of the companies that year after year grow in stature and reputation. They invest in tomorrow rather than feed off today's market. These are companies like Apple where people line up for days in advance to get their hands on the latest and best rather than the new and refined.

This book focuses on the truly creative tools that are efficient and effective to provide an organization with people who are not content with doing things the way they have always been done but who are willing to challenge the system to dream of new and better ways, and then to get busy and make them happen. This is where true self-satisfaction occurs in your people. People who give up vacation because what they are doing at work is more exciting and more rewarding. These are organizations whose stock value continues to increase even when the market is down. These are companies that are held up as the example—the benchmark; they are considered the place to work by people around the world. These are the companies where the people are proud to be part of the excitement and yes, risks, that they live with every day. When you ask these people what they do, they say that there from Apple, or from Google; they do not say they are a programmer, an engineer, or a manager but a member of a team that they are proud of.

These are truly the innovative organizations that are responsible for moving our living environment and living standards forward and upward. These are the companies that really excel. These are the companies whose values and management style could easily be summed up in the following word: *Excelsior.*

We all have two personalities that live inside of us, much as Oscar and Felix featured in the movie, play, and TV show "The Odd Couple." The Felix side of our personality resides in the left-hand hemisphere of our brain. Felix is a well-organized, highly literate individual who loves lists, plans everything he does, and never deviates from the plan. He is driven by rules and the clock. If rules have not been developed, he develops them to define what is expected of him. He likes to have goals either set for him or by himself. He strives to please others and is very disappointed if others do not recognize his efforts.

Oscar, on the other hand, rests in the right-hand hemisphere of our brain, the creative side. Oscar's personality is unstructured, reactive, and driven by whims. He drinks beer for breakfast that was left opened the night before. He challenges authority and rejects conformity. He feels best when he is working on many items, all at the same time. He believes that rules are made to be broken. He marches to his own drummer and relies on self-gratification to keep him going.

Felix had a 4.0 average in college. He loved exams because it proved to his teacher that he did the assignments and learned his lessons. Oscar had a 2.0 average. He created problems in class. He told jokes. He was more interested in making friends than in making grades. Felix works to accomplish something. Oscar functions for the joy of doing it.

We go to school and study to satisfy Felix's needs, to define more rules, to define how things are done and how to plan to accomplish a desired goal. From the time we are born, our parents, then our teachers, then our organizations train us to conform to some predetermined norm. Felix is held up as the example of good. Oscar is the example of bad. Felix always wears a tie, knows what time it is, and always knows what needs to be done next based on past experience or training. Oscar—well, he is out to lunch (see Figure F.1).

Felix uses new rules and regulations to establish a creative idea screen so that he can concentrate his efforts on *getting the job done*. The more rules that he can establish, the finer the filter of the screen, keeping more and more of the creative thoughts from getting through for Felix to consider. The first step in increasing an individual's creativity is to start eliminating rules that screen out creative thoughts (see Figure F.2).

As a result of the feedback we receive throughout our lives, we push our Oscar personality into the shadows, only allowing it to emerge when we relax. Things would work fine if Oscar or Felix lived in the house (our mind) by themselves. But this is not the case. At night when we sleep, Oscar rules with no conflict with Felix because our thoughts do not normally result in accomplishing anything.

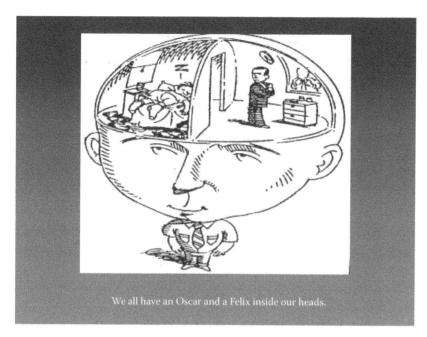

FIGURE F.1

Oscar and Felix sides of our brain.

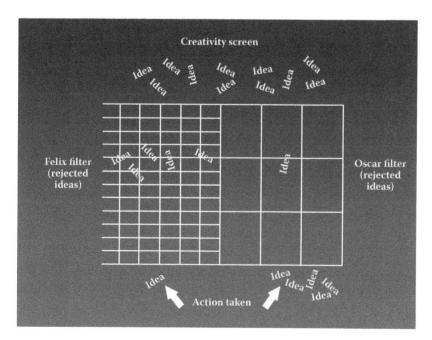

FIGURE F.2

Felix screening out Oscar's ideas.

During the day, both Oscar and Felix live in the same house and must coexist. But Felix rules. Oscar operates like a child, emotionally reacting to outside circumstances. Oscar's outputs are radical and reactive.

As a result, they are processed by Felix before we can put them into action. Felix takes these inputs and tries to put them in order, reshaping and boxing them into Felix's acceptable performance rules. This often results in Felix rejecting the idea and chastising Oscar for even suggesting it.

> Watch your thoughts; they become words. Watch your words; they become actions. Watch your actions; they become habits. Watch your habits; they become character. Watch your character; it becomes your destiny.

Frank Outlaw

Oscar suggests that we should smell the socks and if they do not smell too bad, put them on. Felix reacts by saying, "No way. Rule 1593 states, 'Put clean socks on every day.' How dumb can you be?" Of course, Felix could wind up like the young boy at camp whose mom said to put on clean socks every day. After a week, the boy called his mom and said he could not get any more socks on! He followed the rules but did not get good results.

In other cases, Felix reshapes the idea so drastically that Oscar cannot recognize it. After years of rejection, Oscar just gives up and stops submitting ideas, being content to slip back into the shadow of the individual's dreams. Here are the problems that we face today: How do we get Felix to encourage Oscar to submit ideas? How do we get Felix to react to these ideas in a positive manner as perceived by Oscar? How do we encourage the creative side of our personality to participate more actively in our day-to-day activities? That, simply put, is the objective of this book.

The tools and methodologies that are presented here are the most efficient and effective at driving creativity throughout an organization. The organization can be small, medium, or large. It can be in production, service, or government. It can be American, French, or Chinese. The makeup of the company is not the deciding factor. It can be as big as the US government or as small as the family unit. What really drives creativity is the feeling that rests in each and every individual of excitement, empowerment, trust, integrity, and knowing how to use the right tools and methodologies to allow everyone to dream and then to pursue those dreams.

We are presenting the following tools/methodologies that are most efficient and effective in helping organizations and individuals to develop and produce innovative, creative products, processes, and services, or to improve existing ones.

76 Standard Solutions	Lead user analysis
Absence thinking	Lotus blossom
Biomimicry	Matrix diagram
Combination methods	Mind mapping
Concept tree	Online innovation platforms
Consumer co-creation	Open innovation
Creative thinking	Outcome-driven innovation
Directed/focused/structured innovation	Proactive creativity
Elevator speech	Proof of concept
Ethnography	Quickscore creativity test
Generic creativity tools	Scenario analysis
HU diagrams	Storyboarding
I-TRIZ	Synectics
Imaginary brainstorming	TRIZ
Innovation blueprint	

We are certainly not advocating that you need to use all 29 in order to have a creative environment. Quite the contrary—the innovator needs to understand all 29, how they are used, and the type of results that they bring about in order to select the right combination for the specific organization they are working with.

Preface

In today's fast-moving and high-technology environment, the focus on quality has given way to a focus on innovation. Quality methodology has been shared and integrated into organizations around the world. High quality is now a given for products and services produced in Japan, United States, Germany, Italy, China, India—yes, everywhere. Competition is more fierce and intense than ever before. Technology breakthroughs can be transferred to any part of the world in a matter of days. The people trained in schools around Shanghai are better educated than the people graduating in San Francisco or New York City. The key to being competitive is staying ahead of the competition. That means coming out faster and with more competitive products and services than the competition. The problem that every organization—be it public or private, profit or nonprofit, product or services—has is a need to have more innovative ideas effectively implemented. Today, we need to have our people generate more and better innovative ideas that can be rapidly provided to the consumer. That means that every part of the organization needs to be involved in the innovative activities. Innovative ideas cannot come from just research and development alone. We need to have more innovative processes and systems to support finance, production, sales, marketing, personnel, information technology, procurement—yes, every part of the organization. Even the person sweeping the floors can come up with an innovative idea that will drive a new product cycle. Organizations used to expect their employees, when they came to work, to stop thinking and blindly follow instructions. Today, employees need to realize that they are being paid for both their physical and mental capabilities. Our employees have to understand that if they are going to get ahead, they are now required to be more creative and more innovative at work than any place else. When they get to work, everyone needs to take off their baseball cap and put on their thinking hat in order for the organization to be successful. Today, the best worker is the best thinker, not the one who moves the most products.

Everybody is talking about the importance of innovation, how innovative they are, what innovative products they are producing, and how they need to be more innovative. Everybody is using the word *innovation* to highlight why they are different from everyone else, why customers should rely on them to provide services and product. But what does it all mean? After years of discussion, arguments, and debates, there is little agreement on what a true definition of innovation is. At one extreme, people will argue that innovation is "any new and unique idea." At the other extreme, individuals will define innovation as "a new and unique idea that is produced and delivered to an external customer who is willing to pay more for it than the cost to provide it plus a reasonable profit margin for the supplier." If you use the first definition of innovation, almost all organizations are innovative organizations. If you use the second definition of innovation, less than 5% of organizations could be considered innovative. Over 95% of all new and unique ideas and suggestions that come from within most organizations never see the light of day; in other words, they never reach the final stage of 'deliverables to the customers.' Accordingly, they never generate a profit for the organization. My preferred definition that may or may not be in keeping with your personal beliefs is, "the process of translating an idea or invention into an intangible product, service, or process that creates value for which the consumer (the entity that uses the output from the idea) is willing to pay more for it than the cost to produce it."

Just so there is no confusion between innovation and creativity, creativity is defined as follows:

- *Creative*—Using the ability to make or think of new things involving the process by which new ideas, stories, products, etc., are created.
- *Create*—Make something; to bring something into existence.

The difference between creativity and innovation is that the output from the innovation has to be a value-added output, while the output from creativity does not have to be value added.

Keeping this information in mind, you can grasp the problem that the International Association of Innovation Professionals (IAOIP) was faced with when they were assigned the responsibility for (a) defining the body of knowledge for innovation and (b) establishing a certification program for innovators. It is obvious that before you can establish a body of knowledge for innovation, you have to have an accepted definition of innovation. Moreover, before you can certify an individual as the innovator, you have to be able to define what an innovator is and does.

Again, the definition of an innovator is very debatable. The following are five different definitions of an innovator:

1. An innovator is an individual who creates a unique idea that is marketable.
2. An innovator is an individual or group who creates a unique idea and is able to guide it through the processes necessary to deliver it to the external customer.
3. An innovator is a person who has the capability to create unique ideas and has the entrepreneurship to turn these ideas into output that is marketed.
4. An innovator is a person who creates a unique idea and uses the facilities available to produce an output that is marketable.
5. An innovator is an individual who creates a unique idea that is marketable and guides it through the development process so that its value to the customer is greater than the resources required to produce it.

As you can see, some of the experts in the field define an innovator as "a person who comes up with a unique and creative idea that adds value." Other experts in the field feel that "an innovator must be capable of finding an unfulfilled need and taking it through each phase of the innovative process." This means that the innovator must be capable of

- Defining an unfulfilled need
- Creating a solution for the unfulfilled need
- Developing a value proposition
- Getting the value proposition approved by management, if it relates to an established organization
- Getting the project funded
- Establishing an organization to produce the output
- Producing the output
- Marketing the output
- Selling the output
- Evaluating the success of the project

Now this may be a lot to expect one person to be able to accomplish. But there are literally millions of these individuals successfully doing this today. You can see some excellent examples of these types of individuals on one of the most-watched new TV programs called *Shark Tank*. The contestants are individuals who originated a unique idea, found ways to get it funded, and found ways to produce the output. They also set up the marketing and sales system, and sold the product. These are innovators who have come to the TV program to present their innovative output to a group of five entrepreneurs in order to get additional funding and increase sales opportunities. Basically, we believe an innovator is "a person who identifies an unfulfilled need, creates ideas that will fulfill the unmet need, and incorporate the skills of an entrepreneur."

So, what is an entrepreneur? Is an innovator also the same as an entrepreneur? Basically, an entrepreneur is someone who exercises initiative by organizing a venture to take benefit of an opportunity, and, as the decision maker, decides what, how, and how much of a good or service will be produced. An entrepreneur supplies risk capital as a risk taker, and monitors and controls the business activities. The entrepreneur is usually a sole proprietor, a partner, or the one who owns the majority of shares in an incorporated venture (www.businessdictionary.com).

Keeping this in mind, the real difference between an entrepreneur and an innovator is that the innovator needs to be able to recognize an unfulfilled need and create ideas that fill the unfulfilled need. The entrepreneur does not have to create the idea but can take somebody else's idea and turn it into a value-added output.

As you can see, this discussion applies very nicely to a person who is starting a new organization (e.g., a start-up company), but it is difficult to apply this to an established organization. In an established organization, the innovative process flows through many different functions, and usually the individual who recognized the unfulfilled need and created the idea to fill that unfulfilled need is not assigned to process the idea through the total innovative process. What happens in established companies is that subject matter experts are developed and assigned to individual parts of the innovative process. For example, the controller function is responsible for obtaining adequate financing, marketing defines how the unique concept will be communicated to the customer, and the sales force develops the sales campaign. Manufacturing engineering establishes the production facilities, etc. Each of these functions develops specialized skills to effectively and efficiently process the innovative concept through the innovative cycle. It is a rare exception within an established company when the individual who created the innovative idea is held responsible for having it progress through the innovative cycle and create value added to the consumer and the organization. In the cases where this occurs, that individual is called an entrepreneur. (An intrapreneur is an employee of a large corporation who is given freedom and financial support to create new products, service, systems, etc., and does not have to follow the corporation's usual routines or protocols.) In this book, we will be using the following definition of an innovator:

An innovator is an individual who creates a unique idea that is marketable; one who then guides it through the innovative process so that its value to the customer is greater than the resources required to produce it.

Considering all the difficulties there are in getting an agreed-to definition of innovation or innovator, you might think it would be impossible to define what types of tools and methodologies should be used during the innovative cycle. Not so—we accomplished this by forming a committee to research key documentation related to books and technical articles on innovation methods and techniques. We then prepared a list of recommended tools and methodologies and distributed the list to many of the experts who were teaching or using the innovation process. Each of these experts, in turn, evaluated the long list of tools and methodologies that we had collected, and each put these tools in the following categories:

- Always used
- Frequently used
- Seldom used
- Never used
- Not known

We also asked them to suggest any additional tool/methodology that was missing from the list. As a result of this research study, we defined the tools that were most efficient, effective, and frequently used in the innovative process. These tools are presented in this book along with enough information to show you how to use them. Each tool is represented by a chapter and presented in the following format:

- Definition: tool and/or methodology.
- User: who uses the tool/methodology.
- What phases of the innovative process are the tool/methodology used in?

- How is the tool/methodology used?
- Examples of the outputs from the tool/methodology.
- Software to help in using the tool/methodology.
- References.
- Suggested additional reading.

For many of the tools, there are, or should be, complete books written on how to effectively utilize the tool/methodology. For example, there are a number of books written on storyboarding. In most cases, we have recommended additional reading for those who desire more detailed information on how to effectively implement and use the tool/methodology. We do not believe that most people involved in the innovative process will need to master all of the tools and/or methodologies listed in this book; however, we do believe that all of them are important enough that all the individuals involved in the innovative process should at least be familiar with each of them.

We also recommend that anyone actually involved in moving the innovative idea through any part of the innovative process become active members in the IAOIP. Their mission is to organize and advance innovation through the development of a catalog of innovation skills and capabilities as well as certifications to demonstrate mastery of that body of knowledge. Working groups are organized to create the base of advanced innovation certification requirements and ensure our certified members and organizations are global leaders in practicing and managing innovation. More information related to the IAOIP can be found at http://iaoip.org/.

Acknowledgments

We acknowledge the many hours of work that each of these innovative professionals who wrote chapters for this book expended without compensation. This was done in order to capture and share the knowledge contained in this book. We would be remiss not to acknowledge the many long days and late evenings that Candy Rogers, Joe Mueller, and Susan Koepp-Baker put into proofreading and formatting this book. It was a major challenge to convert the creative thinking of so many individuals into a standard pattern so that the book flowed freely and logically from chapter to chapter.

This book represents the output from the Tools and Methodologies Working Group of the International Association of Innovative Professionals (IAOIP).

About the Editors

 **Dr. H. James Harrington**, chief executive officer (CEO), Harrington Management Systems. In the book *Tech Trending*, Dr. Harrington was referred to as "the quintessential tech trender." The *New York Times* referred to him as having a "... knack for synthesis and an open mind about packaging his knowledge and experience in new ways—characteristics that may matter more as prerequisites for new-economy success than technical wizardry...."

It has been said about him, "Harrington writes the books that other consultants use."

The leading Japanese author on quality, Professor Yoshio Kondo, stated, "Business Process Improvement (methodology) investigated and established by Dr. H. James Harrington and his group bring some of the new strategies which brings revolutionary improvement not only in quality of products and services, but also the business processes which yield the excellent quality of the output."

The father of *total quality control*, Dr. Armand V. Feigenbaum, stated, "Harrington is one of those very rare business leaders who combines outstanding inherent ability, effective management skills, broad technology background and great effectiveness in producing results. His record of accomplishment is a very long, broad, and deep one that is highly and favorably recognized."

Bill Clinton, former president of the United States, appointed Dr. Harrington to serve as an Ambassador of Goodwill.

Newt Gingrich, former Speaker of the House and general chairman of American Solutions, has appointed Dr. Harrington to the advisory board of his Jobs and Prosperity Task Force.

KEY RESPONSIBILITIES

H. James Harrington now serves as the CEO for Harrington Management Systems, and he is on the board of directors for a number of small- to medium-size companies helping them develop their business strategies. He also serves as

- President of the Walter L. Hurd Foundation
- Honorary advisor for quality for China
- Chairman of the Centre for Organizational Excellence Research (COER)
- President of the Altshuller Institute

AWARDS AND RECOGNITION

Harrington received many awards and recognition trophies throughout his 60 years' activity in promoting quality and high performance throughout the world. He has had many performance improvement awards named after him from countries around the world. Some of them are as follows:

- The Harrington/Ishikawa Medal, presented yearly by the Asian Pacific Quality Organization, was named after H. James Harrington to recognize his many contributions to the region.
- The Harrington/Neron Medal was named after H. James Harrington in 1997 for his many contributions to the quality movement in Canada.
- Harrington Best TQM Thesis Award was established in 2004 and named after H. James Harrington by the European Universities Network and e-TQM College.
- Harrington Chair in Performance Excellence was established in 2005 at the Sudan University.
- Harrington Excellence Medal was established in 2007 to recognize an individual who uses the quality tools in a superior manner.
- H. James Harrington Scholarship was established in 2011 by the American Society for Quality Inspection Division.

PUBLICATIONS AND LECTURES

Harrington is the author of more than 40 books and hundreds of papers on performance improvement of which more than 150 have been published in major magazines. He has given hundreds of seminars on every continent except the South Pole.

 **Frank Voehl**, president, Strategy Associates, now serves as the chairman and president of Strategy Associates Inc. and as a senior consultant and chancellor for Harrington Management Systems. He also serves as the chairman of the board for a number of businesses and as a Grand Master Black Belt instructor and technology advisor at the University of Central Florida in Orlando, Florida. He is recognized as one of the world leaders in applying quality measurement and Lean Six Sigma methodologies to business processes.

PREVIOUS EXPERIENCE

Frank Voehl has extensive knowledge of the National Regulatory Commission, the Food and Drug Administration, Good Manufacturing Practice, and the National Aeronautics and Space Administration quality system requirements. He is an expert in ISO-9000, QS-9000/14000/18000, and integrated Lean Six Sigma quality system standards and processes. He earned degrees from St. John's University and advanced studies at New York University, as well as an honorary doctor of divinity degree. Since 1986, he has been responsible for overseeing the implementation of quality management systems with organizations in such diverse industries as telecommunications and utilities, federal, state, and local government agencies, public administration and safety, pharmaceuticals, insurance/banking, manufacturing, and institutes of higher learning. In 2002, he joined the Harrington Group as the chief operating officer (COO) and executive vice president. He has held executive management positions with Florida Power and Light and FPL Group, where he was the founding general manager and COO of QualTec Quality Services for seven years. He has written and published/co-published more than 35 books and hundreds of technical papers on business management, quality improvement, change management, knowledge management, logistics and teambuilding, and has received numerous awards for community leadership, service to third-world countries, and student mentoring.

CREDENTIALS

The Bahamas National Quality Award was developed in 1991 by Voehl to recognize the many contributions of companies in the Caribbean region, and he is an honorary member of its board of judges. In 1980, the City of Yonkers, New York, declared March 7 as *Frank Voehl Day*, honoring him for his many contributions on behalf of thousands of youth in the city where he lived, performed volunteer work, and served as athletic director and coach of the Yonkers-Pelton Basketball Association. In 1985, he was named *Father of the Year* in Broward County, Florida. He also serves as president of the Miami Archdiocesan Council of the St. Vincent de Paul Society, whose mission is to serve the poor and needy throughout South Florida and the world.

About the Contributors

Scott Benjamin is an assistant professor of entrepreneurship and the director for the Center for Entrepreneurship and New Business Development at the Florida Institute of Technology. Before receiving his doctorate in strategic management and entrepreneurship from the Robert H. Smith School of Business at the University of Maryland, Dr. Benjamin had seven start-up ventures in industries ranging from medical product and consulting services to restaurants and real-estate development. His most recent venture, Signature Properties (2001–2013), was a successful real-estate investment and development company specializing in residential and office development. He currently engages with start-up ventures as an active angel investor and works with college students to help them launch their ventures at Florida Tech.

Gül Aslan Damcı has worked as a management and strategy consultant focusing on business strategy, process management, and organizational effectiveness for a variety of industries. She has been engaged in innovation management throughout her career and is currently the chair of the Value Creation Working Group of the International Association of Innovation Professionals. Her specific experience relates to strategy, innovation management, mergers and acquisitions and integration management, process and risk management, and internal audit.

Lisa Friedman, PhD, is a partner in the Enterprise Development Group (EDG), an international consulting and training firm specializing in innovation strategy and best practices. EDG is based in Palo Alto, California, at the heart of Silicon Valley, and works with innovation leaders around the world. EDG works with companies to understand the trends that are revolutionizing their industry, to envision the next generation in their businesses, and to enable a network of leaders and innovators to build this future together (www.enterprisedevelop.com).

Dana J. Landry is a 37-year veteran in medical device design and manufacturing where he held positions in manufacturing, quality, and, for the last 20 years, product research and development. He earned BS, MS, MBA, and PhD degrees with a specialty in leadership studies. He has held certifications as a new product development professional, certified quality engineer, and certified reliability engineer. He is currently an adjunct professor at New York University in the Management of Technology program. Dr. Landry was selected as Outstanding Alumni in Biomedical Engineering in 2012 by Louisiana Tech University for his career dedication to the field. Landry is chairman of the certification committee of the International Association of Innovation Professionals.

Charles Mignosa is president and chief executive officer of Business Management Systems, and vice president and innovation coach with Harrington Management Systems. He is a Lean Six Sigma Master Black Belt, with over 30 years of diversified experience in high technology, telecommunications, food processing and biomedical device industries. He earned a graduate degree in Statistics and Systems Research, and is a senior member of the American Society for Quality (ASQ).

Douglas Nelson is an experienced consultant and manager with international experience in strategic planning, business development, business problem solving, and providing results-oriented management solutions. He is a Lean Six Sigma Master Black Belt/ Process Excellence Professional with proven ability and experience to drive operational process improvement. He has more than 25 years of management and operational experience providing process, cost, and productivity improvement in the manufacturing and service industries.

Jose Carlos Arce Rioboo is the Chief Planning, Development and Innovation Officer of Grupo Rioboo a private own holding of companies, from engineering firms to a k-12 school and a University. He is also partner in Factum-Partners a consulting firm helping companies in strategic planning, project portfolio management and innovation management. He earned a Master in Engineering and an engineering degree from the Universidad Nacional Autónoma de México (UNAM).

Larry R. Smith is presently a senior executive at Harrington Management Systems. He *graduated* (retired) from Ford in 2005 where he was quality coach of the Ford Motor Company's Heritage team, a team charged with revitalizing the Ford Rouge manufacturing complex. Larry's career emphasis has been in quality engineering and product development. He was instrumental in initiating a Six Sigma effort at Ford Motor Company, and acted as deployment champion for the Superduty Truck Platform, whose Six Sigma projects saved $49 million in the first year of operation. He was also champion of a volunteer Ford Environmental Idea Process Team, focused on developing ideas related to design and manufacture of vehicles whose use improves conditions for human beings and the environment. Larry has held a variety of quality management positions in his 27 years at Ford, primarily in vehicle operations, North American truck, powertrain operations, and casting operations. A fellow of the American Society for Quality and elected national director from 2004 to 2006, Smith is a past president of the Altshuller Institute for TRIZ studies and has been involved in numerous applications of TRIZ since 1993.

Maria B. Thompson is the senior director of ITW Global Innovation Framework. She has 30-plus years of experience managing international teams in research, development, invention, and innovation at AT&T Bell Laboratories, Motorola, and ITW. Maria currently facilitates innovation tools workshops to support application of the ITW Global

Innovation Framework. She has more than 20 years of experience working with high-technology teams to generate creative solutions to difficult problems, patent them, and commercialize them.

Brett Trusko is an assistant professor at Texas A&M University and an adjunct assistant professor at Mount Sinai School of Medicine at New York University. He is also the executive director of the Texas Institute for Smart Health and adjunct faculty at Baylor College of Medicine. In this position, he performs research on utilizing innovation, technology, and process improvement to optimize the health-care experience. Dr. Trusko is the founder, president, and CEO of the International Association of Innovation Professionals (IAOIP) and the editor-in-chief of the *International Journal of Innovation Science* (the journal of the IAOIP). He is an author, professor, entrepreneur, speaker, and consultant who works in everything from strategy, to process design to innovation. He splits his time between Manhattan and Houston, Texas. He earned a doctoral degree in business administration.

Stephen P. Walsh is a founding partner of BHW, a successful consultancy formed in 2000 which has helped a considerable number of businesses improve their performance. He is an experienced trainer and consultant in the field of process improvement, including Lean Six Sigma, continuous improvement, and systems thinking. He specializes in transferring the skills of process improvement (or customer satisfaction improvement, as he prefers to call it) to client personnel by combining technical training with train-the-trainer and facilitation skills programs. The continuous improvement approaches developed by Stephen are based on a systems perspective.

Alla Zusman, a TRIZ master, is the director of TRIZ Product Development, Ideation International Inc. TRIZSoft®. She received her TRIZ education from the TRIZ founder Genrich Altshuller and has more than 30 years of TRIZ experience. She was a co-founder of the TRIZ Kishinev School and of the company Progress, the first private engineering company in the former Soviet Union. With more than 30 years of TRIZ experience, Alla is a leader in theoretical and practical work applying the Ideation TRIZ methodology. Mrs. Zusman has made invaluable contributions to the modern development of TRIZ and has both authored and co-authored numerous TRIZ articles and books. She co-authored multiple books (including several with Mr. Altshuller), patents, and numerous papers, and has taught TRIZ to thousands of students around the world.

1

76 Standard Solutions

Dana J. Landry

CONTENTS

DEFINITION

76 Standard Solutions is a collection of problem-solving concepts intended to help innovators develop solutions. A list was developed from referenced works and published in a comparison with the 40 principles to show that those who are familiar with the 40 principles will be able to expand their problem-solving capability. They are grouped into five categories as follows:

1. Improving the system with no or little change: 13 standard solutions
2. Improving the system by changing the system: 23 standard solutions
3. System transitions: 6 standard solutions
4. Detection and measurement: 17 standard solutions
5. Strategies for simplification and improvement: 17 standard solutions

USER

The 76 Standard Solutions can be used effectively by one individual focused on solving a problem or designing a new product or service. It can also be used by a team of individuals working together to solve a problem or to develop a new product, process, or service.

OFTEN USED IN THE FOLLOWING PHASES OF THE INNOVATIVE PROCESS

The following are the seven phases of the innovative cycle. An X after the phase name indicates that the tool/methodology is used during that specific phase.

- Creation phase X
- Value proposition phase X
- Resourcing phase X
- Documentation phase
- Production phase X
- Sales/delivery phase
- Performance analysis phase

TOOL ACTIVITY BY PHASE

- Creation phase—During this phase, the tool may be used to help develop a better understanding of the possibilities of implementing a particular idea. When an idea seems impossible, it can be beneficial to use the 40 Inventive Principles and 76 Standard Solutions to quickly see if maybe there are paths to success that escaped the initial idea review. This analysis effort can help prevent the elimination of an idea before its time.
- Value proposition phase—In this phase, the methodology can help the innovator see uses for a redefined product that may offer a feature

or value proposition not originally considered. It is the process of looking at creation in the variety of new ways forced by the methodology. The astute innovator can often see a new way to satisfy the customer's need, stated or not, or directly affect the *job to be done* by the customer.

- Resourcing phase—While the effect here is not direct, it should not go unmentioned. In financing, either for a corporate project or a venture capital effort, it is beneficial to be able to demonstrate how you arrived at the value proposition. A carefully crafted story that shows how a formal analysis method led to new insights and directly affected the outcome generally has a more beneficial effect than an innovator explaining how much they *know* and why they think the customer will love the new product. Sometimes, some of the 76 Standard Solutions are used to creatively define ways to better utilize the personnel assigned to the project and the other resources that the project will need to utilize.
- Production phase—The production phase is often filled with challenges of how to commercially realize the new invention. These often present themselves as fairly standard engineering problems amenable to standard engineering solutions. If, however, the innovator runs into a seemingly insurmountable problem that endangers the viability of the project, turning to the various TRIZ methods can be beneficial. It is often better to look for a new and better solution than to quickly fall to a lowest common denominator solution that results from looking for quick and easy manufacturing solutions.

HOW THE TOOL IS USED

The following are some key definitions to understand 76 Standard Solutions:

- *TRIZ* is the Anglicized acronym for the Russian phrase for *theory of solving inventive problems*.
- *TRIZ analysis* is a problem-solving methodology based on logic and data to solve problems creatively. As such, TRIZ brings repeatability, predictability, and reliability to the problem-solving process with its structured and algorithmic approach.

- *Substance–field (Su–field) analysis* is a TRIZ analytical tool for modeling problems related to existing technological systems.
- *ARIZ* is an evolutionary development of the TRIZ methodology.
- *I-TRIZ* is a software suite of applications that takes Altshuller's TRIZ 40 principles of conflict resolution and combines them with his 39 characteristics of technical systems. Once the opportunity description is entered into the software package, it analyzes the problem and proposes a number of approaches to improving the design or solving a problem.

Putting the 76 Standard Solutions to Work

In use, the 76 Standards Solutions would be employed after the innovator has become well versed in the 40 principles and learned the concept of Su–field analysis. The 76 Standard Solutions can be used by innovators to significantly improve existing systems, for example, by resolving a system contradiction through the introduction of some entirely new element. Examples include replacing the standard transmission of a car with an automatic transmission or replacing rearview mirrors with rearview cameras.

Typically, the 76 Standard Solutions are used as a step in ARIZ, after the Su–field model has been developed and any constraints on the solution have been identified (Terninko et al., 2000). ARIZ is an evolutionary development to the TRIZ methodology and forms a core to the analytical methods. This aspect of the methodology provides specific sequential steps for developing a solution for complex problems. The first version of ARIZ was developed in 1968 and underwent many modifications during the next 20 years, during which it became a more precise tool for solving technical problems.

The potential user of the methods that fall within the 76 Standard Solutions and the body of tools that are known under the TRIZ or ARIZ monikers should seek specialized training from those experts in both teaching and applying the methodology. There are firms that claim to be direct student descendants from the inventor of the TRIZ methods and other firms that are second generation in nature. The innovator is advised to research potential firms carefully and try to find one with expertise of applying the methods in the field of interest or one close to it.

The methodology is heavily focused on technology and technical problem solving. As such, its use in services, business model innovations, and similar potential areas for innovation is not clear at this time. Innovators focused on these areas are encouraged to seek more appropriate tools until enough data have been created to demonstrate efficient and effective applications of the 76 Standard Solutions to nontechnical fields.

Most proponents of the TRIZ methodology will make the case that methods such as the 76 Standard Solutions can be applied to almost any problem at any time in the product development process. While this may well be true, the most appropriate times for use by the innovator, as opposed to the problem solver working on a current product, is generally early in the product development process, when the innovator is looking for new solutions that might bring a new value proposition to the customer, and during the early concept and prototype development stages, when the innovator often runs into technical roadblocks and the methodology can greatly assist in driving the innovator toward an *inventive* solution.

To bring a vision to the 76 Standard Solutions, they are often presented in a table compared to the 40 Inventive Principles as presented in the *Global Innovation Science Handbook* (1st Ed.) and shown in Table 1.1. For example, the first principle presented is the principle of segmenting the problem. Among the 76 Standard Solutions, there are three most appropriate to the approach, as shown to the right of the specific principle. The table continues through all 40 Inventive Principles. A way to look at this from the innovator's point of view is, when faced with a problem, try this—Segmentation—and do it this way: (a) divide the element, (b) use particles, and (c) divide the article first into parts and then make it flexible by linking the parts. You will note that not all of the 40 Inventive Principles are directly correlated to a specific solution. This does not mean there are no solutions; it just means that they will be found using the full range of TRIZ/ARIZ methodology. The application of this to technology and things is obvious. The applicability of these solutions to other forms of innovation, especially the innovation of business models, remains to be demonstrated.

TABLE 1.1

Mapping the 76 Standard Solutions onto the 40 Inventive Principles in TRIZ

	40 Inventive Principles	76 Standard Solutions
1	Segmentation	Divide the element into smaller units.
		Use particles instead of the whole object.
		Divide the object into parts, then make it flexible by linking the parts.
		Transition to the micro-level.
2	Take out	
3	Local quality	Protect certain regions from the full impact of an action.
		Turn a magnetic field on or off according to the local need.
		Change from uniform structure to a structure that is specific to the situation.
		Concentrate an additive in one location.
4	Asymmetry	Change from uniform structure to a structure that is specific to the situation.
5	Merging	Additive, temporary or permanent, internal or external, from the environment or from changing the environment.
		Simplification of bi- and poly-systems.
6	Universality	
7	Nested doll	
8	Antiweight	
9	Preliminary antiaction	
10	Preliminary action	
11	Cushion in advance	Use a substance to protect a weaker substance from a potentially harmful occurrence.
12	Equipotentiality	
13	Other way around	Introduce magnetic materials in the environment, instead of into the object.
14	Spheroidality or use of curves	
15	Dynamism	Make the system flexible.
		Use dynamic magnetic fields.
16	Partial/excessive action	Control small quantities by applying and removing a surplus.
		Simulate the introduction of more than is acceptable.
17	Another dimension	

(*Continued*)

TABLE 1.1 (CONTINUED)

Mapping the 76 Standard Solutions onto the 40 Inventive Principles in TRIZ

	40 Inventive Principles	76 Standard Solutions
18	Mechanical vibration	Match the natural frequencies of the field with the substance.
		Use vibration in conjunction with magnetic fields.
		Measure changes in a system by means of changes in its resonant frequency.
19	Periodic action	Replace an uncontrolled field with a structured one.
		Use magnetic field resonance.
20	Continuity of action	Do one operation during the downtime of another.
21	Skipping (do fast)	
22	Blessing in disguise	Eliminate harmful effects.
23	Feedback	Self-controlled changes.
		Use dynamic magnetic fields.
24	Intermediary	Use one object to make the actions of another possible.
		Create structures by use of magnetic particles.
		Introduce a ferromagnetic additive, temporarily.
		Use a temporary additive, internal or external.
		Introduce an additive temporarily.
		Measure a copy.
25	Self-service	Self-controlled changes.
		Use dynamic magnetic fields.
26	Copying	Measure a copy.
		Apply additives to a copy instead of the original.
27	Cheap short life	
28	Replace mechanical system with fields	Replace or supplement a poorly controlled field with a more easily controlled field.
		Use of ferromagnetism and ferromagnetic materials.
		Use electric current instead of magnetic particles.
		Create a field that can be detected or measured.
		Use a field instead of a substance.
29	Pneumatic/hydraulic	Use magnetic liquids.
		Use *nothing*.
		Use *nothing* to simulate structures.
30	Flexible shell, films	Change from a uniform structure to a structure that is specific to the situation.

(*Continued*)

TABLE 1.1 (CONTINUED)

Mapping the 76 Standard Solutions onto the 40 Inventive Principles in TRIZ

	40 Inventive Principles	76 Standard Solutions
31	Porous materials	Use porous or capillary materials.
		Change a uniform structure to a nonuniform one.
		Use capillary or porous structures in a magnetic material, or to contain magnetic fluid.
32	Change color	Use detection instead of measurement.
		Measure the system by means of natural phenomena.
33	Homogeneity	
34	Discard/recover	The additive disappears after use.
35	Change parameters	Phase change.
		Additive, temporary or permanent, internal or external, from the environment or changing the environment.
		Use rheological liquids.
36	Use phase transition	Use the accompanying effects from phase changes.
		Use the physical effects of magnetic transitions.
		Control a system by means of a phase transition, instead of measuring temperature, pressure, magnetic field, etc.
		Measure the system by means of natural phenomena.
37	Use thermal expansion	Control a system by means of thermal expansion, instead of measuring temperature.
		Measure expansion instead of temperature.
38	Strong oxidants	Getting needed ions, molecules, etc.
		Use small amounts of very active additives.

Example

For examples of TRIZ applications, we recommend you visit the Altshuller Institute website (https://www.aitriz.org/). Any presentation of one or two examples would be an injustice to the methodology, which, when used by a properly trained and motivated innovator, can reap considerable benefits in insight, problem solving, and new ways of moving forward with development of one's ideas and inventions.

SOFTWARE

Software to support the 76 Standard Solutions is most often found under the title "TRIZ Software." This will generally come in two main forms. The first is software designed by firms that also sell TRIZ consulting services. The second are free sites, often set up by professors who teach this subject, to help students use the methodology and generate solutions to class or real-life problems.

It is not absolutely necessary to employ software to use the methodology, but experience has shown that for larger or more complex problems, it may offer significant advantages in assuring that all possible steps are utilized, that no potential path is forgotten or missed, and that the innovator can document the analysis in order to build new ones and preserve the knowledge gained.

REFERENCE

Terninko, J., Domb, E., and Miller, J. *The Seventy Six Standard Solutions, with Examples Section One.* Downloaded from http://www.triz-journal.com/seventy-six-standard-solutions-examples-section-one/, February 26, 2000.

2

Absence Thinking

Frank Voehl

CONTENTS

DEFINITION

Absence thinking involves training the mind to think creatively about what it is thinking and not thinking. When you are thinking about a specific something, you often notice what is not there, you watch what people are not doing, and you make lists of things that you normally forget. It involves trying to deliberately think about what is absent and envisioning what you are not thinking about.

USER

This tool can be used by individuals or groups, but its best use is with a group of four to eight people. Cross-functional teams usually yield the best results from this activity. Since absence thinking is a creative method to find new ideas to solve problems and to find innovative products and services, this method can be very usefully combined with a brainstorming session. The core idea is for a team to forget everything they have learned and retained when thinking about a possible solution—that is, wipe the slate clean. Both individuals and groups can use it when stuck and unable to shift thinking to other modes.

OFTEN USED IN THE FOLLOWING PHASES OF THE INNOVATIVE PROCESS

The following are the seven phases of the innovative cycle. An X after the phase name indicates that the tool/methodology is used during that specific phase.

- Creation phase X
- Value proposition phase X
- Resourcing phase
- Documentation phase
- Production phase
- Sales/delivery phase
- Performance analysis phase

TOOL ACTIVITY BY PHASE

- Creation and value proposition phases—This tool is primarily used during the creative and early value proposition phases. It is used to focus on potential project risks and to provide new and different alternatives to complex problems.

HOW TO USE IT

The psychology of thought is such that we are very good at seeing what is there, but not at all good at seeing what is not there. Absence thinking compensates for this by deliberately forcing us to do what we do not naturally do. The images you summon up with absence thinking have an individual structure that may indicate an underlying idea or theme. Your unconscious mind is trying to communicate something specific to you, though it may not be immediately comprehensible. The images can be used as armatures on which to hang new relationships and associations.

Absence thinking involves training your mind to think creatively about what you are thinking and to also think about what you are not thinking; sort of a *ying and yang* technique. When you are thinking about a specific thing, you often notice what is not there, you watch people and what they are not doing, and you make lists of things that you normally forget. In other words, you try to deliberately think about what is absent and envision *what is not there*. Both individuals and groups can use it when stuck and unable to shift thinking to other modes. Also, you can use it when you want to do something that has not been done before.

The following steps outline how to use the absence thinking technique:

1. Divide a piece of paper in half and write down what you are thinking about on the left side, and then write down what you are not thinking about on the right side.
2. Next, when you are looking at something (or otherwise sensing), notice what is *not* there.
3. You can also watch people and notice what they do not do.
4. Some use it to make lists of things that they normally forget in order to remember them.

To be really useful, deliberately and carefully think about what is absent:

- Use it when you are stuck and unable to shift thinking to other modes.
- Use it when you want to do something that has not been done before.

- Think about what you are thinking about, and then think about what you are not thinking about.
- When you are looking at something (or otherwise sensing), notice what is *not* there.
- Watch people and notice what they do not do.
- Make lists of things to remember that you normally forget.
- In other words, deliberately and carefully think about what is absent.

The following is Michael Michalko's detailed step-by-step blueprint for using this technique*:

1. Think about your challenge. Consider your progress, your obstacles, your alternatives, and so on. Then push it away and relax.
2. Totally relax your body. Sit on a chair. Hold a spoon loosely in one of your hands over a plate. Try to achieve the deepest muscle relaxation you can.
3. Quiet your mind. Do not think of what went on during the day or your challenges and problems. Clear your mind of chatter.
4. Quiet your eyes. You cannot look for these images. Be passive. You need to achieve a total absence of any kind of voluntary attention. Become helpless and involuntary and directionless. You can enter the hypnogogic state this way, and, should you begin to fall asleep, you will drop the spoon and awaken in time to capture the images.
5. Record your experiences immediately after they occur. The images will be mixed and unexpected, and will recede rapidly. They could be patterns, clouds of colors, or objects.
6. Look for the associative link. Write down the first things that occur to you after your experience.
7. Look for links and connections to your challenge. Ask questions such as
   - What puzzles me?
   - Is there any relationship to the challenge?
   - Any new insights? Messages?

---

* Michael Michalko is the author of the highly acclaimed *Thinkertoys: A Handbook of Creative Thinking Techniques; Cracking Creativity: The Secrets of Creative Genius; ThinkPak: A Brainstorming Card Deck* and *Creative Thinkering: Putting Your Imagination to Work.* http://creative thinking.net/WP01_Home.htm. See more details at: http://creativethinking.net/#sthash.uy06CDjc .dpbs.

- What is out of place?
- What disturbs me?
- What do the images remind me of?
- What are the similarities?
- What analogies can I make?
- What associations can I make?
- How do the images represent the solution to the problem?

For more detail, see the following website link: http://www.creativity post.com/create/salvador_dalis_creative_thinking_technique#sthash .R5R0aDin.dpuf.

Advantages

Every new product, program, or service results from an idea that then follows an innovation cycle of testing, implementing, and marketing. It is generally accepted that the key to competitive advantage is generating and successfully exploiting new ideas that come from creative thinking of both individuals and groups, and absence thinking can help stimulate new ideas and force us to do what does not come naturally.

Disadvantages

The environment in some organizations may prove to be hostile to creative thinking. Anyone managing this creative thinking process is likely to encounter obstacles such as (a) general resistance to change and lack of comfort with what does not come naturally to us; (b) free expression being stifled by a pervasive culture of blame; (c) rigid formalities and rules that discourage techniques like absence thinking; (d) inadequate and nonexistent incentives leading to slow new ways of thinking; (e) reluctance to think and move outside of strict job categories and ways of thinking; and (f) a view that the best ideas come from the normal ways of thinking, and not from voids and *nonthinking patterns.*

Summary

This technique relies on the fact that people are very good at seeing what is there, but not at all good at seeing what is not there. Absence thinking compensates for this by deliberately forcing us to notice things that are not usually apparent. Think about your challenge, consider your progress, your obstacles, your alternatives, and so on. Then push it away and relax. For example, watch people and notice what they do not do. Make lists of things to remember, those that you normally forget. In other words, deliberately and carefully think about what is absent.

This activity is helpful when you are stuck and unable to shift thinking to some other approach. It is analogous to the importance of negative space to artists. For writers, it may help shift perspective in a story from foreground to background or from the view of a central to a peripheral character or event. For writing contracts, it forces you to imagine what your client will question or what trap might kill a deal.

The psychology of the creative thinking process is such that while we are very good at seeing what is there, we need to do a better job at seeing what is not there. Absence thinking compensates for this by deliberately forcing us to do what do not naturally come to us.

EXAMPLES USING VARIOUS INNOVATIVE TYPES OF ABSENCE THINKING

Salvador Dali Absence Thinking Method: The Tin Plate:

> Dali was intrigued with the images that occur at the boundary between sleeping and waking. They can occur when people are falling asleep, or when they are starting to wake up, and they tend to be extremely vivid, colorful, and bizarre. He experimented with various ways of generating and capturing these fantastical images. His favorite technique is that he would put a tin plate on the floor and then sit by a chair beside it, holding a spoon over the plate. He would then totally relax his body; sometimes he would begin to fall asleep. The moment that he began to doze, the spoon would slip from his fingers and clang on the plate, immediately waking him to capture the surreal images. The absence thinking images seemed to appear from nowhere, but there was a logic in it for Dali. He reasoned that his unconscious is a living, moving stream of energy from which absence thinking gradually rises to the conscious level and takes on a definite form. He then postulated that your unconscious is like a hydrant in the yard, while your consciousness is like a faucet upstairs in the house. Once you know how to turn on the hydrant using absence thinking, a constant supply of images

can flow freely from the faucet. These forms give rise to new thoughts as you interpret the strange conjunctions and chance combinations. See more at http://www.creativitypost.com/create/salvador_dalis_creative_thinking _technique#sthash.R5R0aDin.dpuf.

1. A restaurant owner uses the Dali Tin Plate technique to inspire new promotion ideas.

When the noise awakened him, he kept seeing giant neon images of different foods: neon ice cream, neon pickles, neon chips, neon coffee, and so on. The associative link he saw between the absence of various foods and his challenge was to somehow use the food itself as a promotion. The idea he came up with was to offer various free food items according to the day of week, the time of day, and the season.

For instance, he might offer free pickles on Monday, free ice cream between 2 and 4 p.m. on Tuesdays, free coffee on Wednesday nights, free sweet rolls on Friday mornings, free salads between 6 and 8 p.m. on Saturdays, and so on. He advertises the free food items with neon signs, but you will never know what food items are being offered free until you go into the restaurant.

The sheer variety of free items and the intriguing way in which they are offered has made his restaurant a popular place to eat. Another promotion he created as a result of using absence thinking to visualize images of different foods is a frequent-eater program. Anyone who hosts five meals in a calendar month gets $30 worth of free meals. The minimum bill is $20 but he says the average is $30 a head. These two promotions have made him a success. See more at http://www.creativitypost.com/create/salvador_dalis_creative_thinking _technique#sthash.R5R0aDin.dpuf.

2. An artist draws the spaces between things instead of allowing the spaces to occur naturally, thereby stimulating new ways of visualizing.

3. An R&D manager for a furniture manufacturer considers product areas where customers have made no comment. She observes them carefully using tables, and notes that they leave the tables in plain sight and unattended when not using them. She then invents a table that can be easily be folded and stored.

SOFTWARE

- *iThinkerBoard for the iPad and ThinkPack for the card-deck*: Looking for a unique invention, an untapped market for an existing product, or a new solution? Stretch and flex your mental muscles with iThinkerBoard and a card-deck version, ThinkPak—creative thinking tools that provide a unique way to organize and visualize your ideas. You can use them for absence thinking as a life planner, project management tool, or task organizer. ThinkPak is 56 individual cards used to create new and innovative ideas. Not only can the cards be used individually but also with groups, co-workers, teammates, family, children, etc. If you have not read the books *Thinkertoys*, or *Cracking Creativity*, these tools are merely a way to create new ideas.
- *The Creative Thinker by Idon Resources*: This software brings you nonlinear, yet structured, thinking with the ability to further develop your thoughts in the form of user-friendly and fun-to-use graphical hexagon modeling for absence thinking. You can even organize your knowledge, supplement ideas with notes, and directly and meaningfully link ideas to the Internet, documents, slide presentations, spreadsheets, and more. The Creative Thinker literally allows you to visualize the gaps in your thoughts, rapidly access knowledge, and combine the two creatively in real time for insights. See http://www .idonresources.com/ct/creativethinker.html.
- *Creative Thinking program applied to absence thinking*: The objective is to use the absence listing tool in combination with other innovation methods to drive innovation, improvization, and creative thinking into a team and organization. A combination of skill development training, real-time process facilitation, team and individual coaching, and interviews are usually used by the innovation coach. Programs include a repeatable creative thinking for attributes process that can be used by the different teams, along with relative thinking best practices, tools, and techniques for individuals and teams. See http://www.creativeemergence.com /creativethinking.html.

REFERENCE

Michalko, M. *Thinkertoys: A Handbook of Creative Thinking Techniques; Cracking Creativity: The Secrets of Creative Genius; ThinkPak: A Brainstorming Card Deck and Creative Thinkering: Putting Your Imagination to Work.* Battleboro, VT: Prince. http://creativethinking.net/#sthash.uy06CDjc.dpbs.

SUGGESTED ADDITIONAL READING

Higgins, J. *101 Creative Problem Solving Techniques: The Handbook of New Ideas for Business.* Revised edition. Winter Park, FL: New Management Pub Co., 2005.
Mauzy, J. and Harriman, R. *Creativity Inc.: Building an Inventive Organization.* Boston: Harvard Business School, 2003.
Prince, G.M. *The Practice of Creativity: A Manual for Dynamic Group Problem-Solving.* Vermont: Echo Point Books & Media, LLC, 2012. 0-9638-7848-4.
Roth, W. and Voehl, F. *Problem Solving for Results.* Delray, FL: St Lucie Press, 1998.

3

Biomimicry (Also Known as Biomimetics, Biogenesis, Biognosis)

Dana J. Landry

CONTENTS

DEFINITION

Biomimetic or biomimicry is the imitation of the models, systems, and elements of nature for the purpose of solving complex human problems (Wikipedia). It is the transfer of ideas from biology to technology, the design and production of materials, structures, and systems that are modeled on biological entities and processes. The process involves understanding a problem and observational capability together with the capacity to synthesize different observations into a vision for solving a problem.

USERS

This tool is most effectively used with small groups of individuals (three to eight people) who are working on complex problems (Benyus, 2009).

OFTEN USED IN THE FOLLOWING PHASES OF THE INNOVATIVE PROCESS

The following are the seven phases of the innovative cycle. An X after the phase name indicates that the tool/methodology is used during that specific phase.

- Creation phase X
- Value proposition phase
- Resourcing phase
- Documentation phase
- Production phase X
- Sales/delivery phase
- Performance analysis phase

TOOL ACTIVITY BY PHASE

- Creative phase—During this phase, biomimicry is used to transfer thought patterns into physical terms that are used within the organization (specifications, procedures, processes, processes, prints, etc.).
- Production phase—During this phase, the tool is primarily used to help define different approaches to effectively producing the product.

HOW TO USE THE TOOL

- Begin with the question, "How would nature solve this problem?"
- Teams can include a biologist early in the design phase or can consult databases of *nature solutions* such as AskNature.org, which catalogs how nature solves problems.
- The team then models a solution to the problem on the basis of the inputs from nature.
- The team proceeds to develop the design using typical methods.

EXAMPLES OF THE TOOL OUTPUT

- Velcro was modeled after the spiny hooks on plant seeds and fruits.
- The Japanese bullet train front end is based on the kingfisher beak design as it allows the bird to enter water with minimal splash, allowing the bullet train to cut through the atmosphere with minimum air disturbance.
- People study how animals attach themselves to objects while underwater to develop glues that will cure underwater or wet.
- Superefficient fan blades, aerators, and propellers have been designed on the basis of the geometry of the flow-friendly spiral found in seashells, kelp, and ram horns.

SOFTWARE

No specific software related to this tool is recommended.

REFERENCE

Benyus, J.M. *Biomimicry: Innovation Inspired by Nature.* New York: HarperCollins, 2009.

SUGGESTED ADDITIONAL READING

Lakhtakia, A. and Martin-Palma, R.J., eds. *Engineered Biomimicry.* Amsterdam: Elsevier, 2013.

4

Combination Methods

Gül Aslan Damcı

CONTENTS

DEFINITION

The combination method is a by-product of already applied process, system, product, and service wise solutions integrated into a one solution system to produce one end result that is unique.

USER

Teams that are cross-functional as well as composed of individuals from different industries or different expertise would provide ideal platforms for both the selection and implementation of combined methods.

25

OFTEN USED IN THE FOLLOWING PHASES OF THE INNOVATIVE PROCESS

The following are the seven phases of the innovative cycle. An X after the phase name indicates that the tool/methodology is used during that specific phase.

- Creation phase X
- Value proposition phase
- Resourcing phase
- Documentation phase
- Production phase X
- Sales/delivery phase X
- Performance analysis phase

TOOL ACTIVITY BY PHASE

- Creation phase—During this phase, the combination method approach can effectively be used by combining together a number of current products, thereby creating a new product that is capable of performing all the tasks that the combination of things can do in a smaller and less expensive package.
- Production phase—During the production phase, the combination method can be used to bring together activities that were done on numerous pieces of equipment, so that they can be done on a single piece of equipment faster and in a less expensive way.
- Sales/delivery phase—During this phase, the combination method can be used to combine different marketing or sales techniques together. For example, an advertisement used on television could also be used on YouTube.

HOW TO USE THE TOOL

The combination method is a technique that combines various innovative processes (at least two) for creating a new product or service, or for problem solving. As such, there are two different categories of combination methods.

- Category I—combining innovative processes that would separate or collectively be related to structures, functions, principles, processes, and technologies. In more detail, this method is about combining already existing products, features, production processes, and functions to develop a new product or service.
- Category II—combining methods of problem solving that require specialized solutions. In fact, the approach may yield good results for problem solving through the application of different techniques with distinct approaches to the problem. The method combines already accepted problem-solving methodologies.

There are general techniques implemented for combined solutions.

Category I: Combination of Existing Outcomes/Solutions

- Combination of production processes/techniques
- Combination of materials
- Combination of end products (including by-products)
- Combination of products and services

The generation and development of combination methods would be the process of searching for best outcomes to provide one solution manifested as an end process, solution, product, and service. The best combination method would comprise one or a combination of the following: integrations of materials, production techniques, functions/features, design, etc. As shown by real cases that will be presented below, the implementation and research and development of the combination method requires certain aspects. In summary, approaches and revisions to business models, determination and use of right competences and knowledge, identification of right process requirements, and adoption into own processes will be required for the adoption of combined methods.

The usage of combination methods can be clustered into six areas that are presented as follows:

1. Combination of technologies
   This technique would be the combination of at least two different technology solutions to produce an end result. The wine press and

coin punch technologies inspired and were used by Gutenberg when he invented the printing press in 1439.

2. Combination of materials

A new material is produced with the combination of two different materials. The new material would have unique characteristics that would allow for different areas of usage. The combination of copper and tin allowed for the discovery of bronze around the third millennium BC, which allowed for metal objects to be more durable and harder.

3. Combination of products

At least two or more products are combined to produce a new product. Owing to the already existing and known traits of the combined products, the end product would allow for additional functions embedded into a single product and would not necessarily be as unique as the other techniques. The distinctiveness would be about the ease that a single product delivers with additional functions. All-in-one products, such as remote controllers used for controlling several devices at home (e.g., television, air conditioner, music set) are some examples.

4. Combination of functions and structures

This technique is about adding extra functions to an existing product or combining different functions to produce a new product. The end product will have distinct characteristics in terms of functionality and areas of use. The technique would aim to fill the gap of unmet needs of customers by fulfilling additional needs with the same product and service. For example, air conditioners and heaters developed as artwork are good illustrations of meeting unmet needs and wants. The production process and material use will not differ. Introduction of modular heating systems by Joris Laarman and Jaga as pieces of art, such as bas-relief hangs, represent how art and function can be combined. The inflatable sleeping coat is an example that is a Red Dot Award winner for design innovation.

5. Combination of structures

The technique is about combining different structures. A Fiat automobile factory with the rooftop being used as a test track is a combined structure example.

6. Combination of design

The combination will be the use of a design for an already existing product for a different product marketed to meet different needs and wants of customers. The already well-known design will be infused to another product and would embed the familiarity and likeness aspect. LEGO uses different designs of objects within new LEGO series. At the same time, LEGO-shaped couches and other accessories (e.g., pen, scissors) are good examples of brand extensions.

Cross-industry innovation is a good example of the combination method, which is defined as the "process by which firms systematically incorporate external knowledge, concepts or technologies from more or less distant industries into their own innovation process" (Bader et al., 2013). Collaboration across industries is the key dynamic and is actually a direct example of the application of the combination method.

Category II: Combination of Problem-Solving Methodologies

Depending on the nature of the problem, widely used problem-solving methodologies for specific problem categories may be combined. The combination technique would be about taking the relevant problem-solving approaches of each methodology and creating a combined approach to analyze and solve the problem. There are well-known problem-solving techniques that are individually effective methods. The following list presents the methodologies that are mostly used for problem solving:

   i. Six Sigma
  ii. Root cause analysis
 iii. Lean
 iv. TRIZ method

One key advantage of the techniques is that the problem solver will not be limited with a single problem-solving framework, and will integrate different approaches from other techniques at the required problem-solving stages.

Benefit

Uncertainty and risk and relatively lower through innovation by the use of the combination method, as the solutions have already been tried and implemented for other products, services, and in different industries. The combined method facilitates the development and improvement of existing business models through the transfer of different business model components of various industries.

EXAMPLE

3M is one global corporation that has been very successful with radical innovations such as the Littmann electronic stethoscope or Post-it notes. 3M devotes a fundamental portion of innovation efforts to *inside–out* and *outside–in* cross-industry processes.

3M uses the know-how of lead users in other industries and transfers these to their own processes. The lead user methodology was indeed very successful in developing a revolutionary method for the surgical infection control field. The lead users for the area were theater makeup artists, veterinary scientists, and oceanographers who have all contributed to the development of the antimicrobial armor line. The lead user methodology stands as the *outside–in* cross-industry innovation.

The inside–out process is about the transfer of a specific solution across industries. 3M's use of nonwoven fabrics is a transfer of their use in automotive and chemical industries. Some materials can be used for multiple purposes, as well as in different markets with the same use. This example represents the use of combined materials for a different functional use.

Aesculap is another global health-care company specialized in surgical medicine that has incorporated cross-industry innovation into its business model. Aesculap involves external parties specialized in biomaterials and computer hardware for improving products and services. Aesculap used cross-industry innovation for the development of the Procedure Kit product and service, where it transferred fleet management process for the use of Procedure Kit for highly complex surgeries. Insights from another industry were integrated for the development of this business model that represents a combined method.

Table 4.1 provides more examples of cross-industry innovation.

TABLE 4.1

Cross-Industry Examples from the Health-Care Sector

Company	Cross-Industry Innovation Examples
3M	Medical tapes, electronic stethoscopes, surgical masks
MedCorp2	Use of fleet management idea from automotive industry for surgical instruments for hospitals and on protective equipment for chemical plants
Aesculap	Use of fleet management idea from automotive industry for surgical instruments
MedCorp4	Improved processes via the transfer of air cargo processes and services and implement on treatments for patients
MedCorp5	Transferring a solution from the automotive industry for a lighting system
MedCorp7	Bone cement production through cross-business-unit cooperation

There are several common themes drawn from the cases presented in Table 4.1. These corporations have used one or more of the following processes to realize cross-industry innovation and come up with successful results. The key techniques that these corporations have used are summarized as follows:

1. Integration of main lead users from different industries
2. Innovation via cross–business unit innovation
3. Transfer of different industry business models into own industry (e.g., transfer of fleet management idea from automotive industry to the health-care sector)
4. Implementation of a single solution or technology in different market domains (e.g., internal improvement of different production processes by transfer of know-how across business sectors)

SOFTWARE

No particular software is recommended for this tool.

REFERENCE

Bader, K., Buchholz, C., Bohn, L., and Enkel, E. A view beyond the horizon: Cross-industry innovation. *Performance*, vol. 5, no. 2, May 2013.

SUGGESTED ADDITIONAL READING

Conley, D. Innovation combination methods. In: P. Gupta and B.E. Trusko, eds. *Global Innovation Science Handbook*. New York: McGraw-Hill Education, 2014.

Gassmann, O. and Zeschky, M. Opening up the solution space: The role of analogical thinking for breakthrough product innovation. *Creativity and Innovation Management*, vol. 17, no. 2, pp. 97–106, 2008.

Kalogerakis, K., Lüthje, C., and Herstatt, C. Developing innovations based on analogies: Experience from design and engineering consultants. *Journal of Product Innovation Management*, vol. 27, no. 3, pp. 418–436, 2010.

WEBSITES

http://www.slideshare.net/marcnewshoestoday/37-ways-for-innovation-by-combination
http://www.triz-journal.com/comparison-innovation-methodologies-triz/

5

Concept Tree (Clustering)

Charles Mignosa

CONTENTS

DEFINITION

Conceptual clustering is the inherent structure of the data (concepts) that drives cluster formation. Since this technique is dependent on language, it is open to interpretation and consensus is required.

OFTEN USED IN THE FOLLOWING PHASES OF THE INNOVATIVE PROCESS

The following are the seven phases of the innovative cycle. An X after the phase name indicates that the tool/methodology is used during that specific phase.

- Creation phase X
- Value proposition phase

- Resourcing phase
- Documentation phase
- Production phase X
- Sales/delivery phase
- Performance analysis phase

TOOL ACTIVITY BY PHASE

This tool is primarily used during the creative and production phases. Because it is used to help analyze problems, it could be used in any phase when a problem occurred.

HOW THE TOOL IS USED

A concept tree is a powerful, visual way to leverage current ideas to generate dozens of others, and find a unique solution to your challenge. It is also useful if your original idea is too general, has some limitations, or is not actionable. A sample of a concept map is shown in Figure 5.1.

- Identify the highest-level definition of the area or process of interest.
- Identify each lower level reporting to the above level.
- Construct the chart by attaching each lower level to the appropriate upper level.

Conceptual trees come in many forms depending on the subject being evaluated.

Typically, the tree starts at the highest level and drills down to the individual components and the lowest level. Example: a planet, the Earth: water areas, land, continents, vegetation, animals, *Homo sapiens*, male/female (Figure 5.1).

A decision tree is another kind of concept tree that involves activities that lead to a final result. Making breakfast with tea, eggs, and toast might look like Figure 5.2.

Figure 5.3 is an educational concept tree.

Figure 5.4 is a concept tree for a monetary grants system.

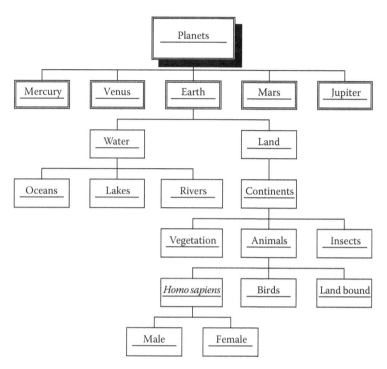

FIGURE 5.1
Concept tree of a plan broken down to its lower entities.

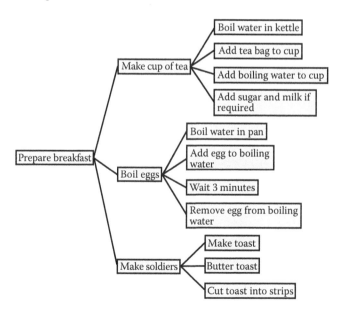

FIGURE 5.2
Typical decision tree. (From http://www.epistemics.co.uk.)

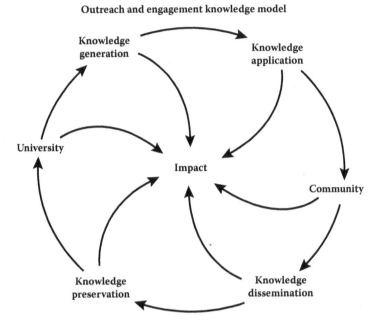

FIGURE 5.3
Educational concept tree. (From *Engaged Scholar*, Michigan State University, http:// engagedscholar.msu.edu.)

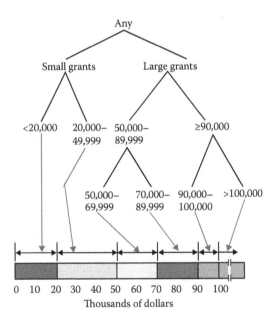

FIGURE 5.4
Concept tree for monetary grants. (Data cubes from http://www2.cs.uregina.ca.)

Basic Steps in Constructing a Concept Tree

- Identify the highest-level definition of the area or process of interest.
- Identify each lower level reporting to the above level.
- Construct the chart by attaching each lower level to the appropriate upper level.

EXAMPLE

Some examples of this tool are included in the section of this chapter entitled "How the Tool Is Used."

SOFTWARE

Some commercial software available includes but is not limited to

- MindView 5: http://mindmappingsoftwareblog.com
- NovaMind: http://mindmappingsoftwareblog.com
- XMind: http://mindmappingsoftwareblog.com

SUGGESTED ADDITIONAL READING

Biswas, G., Weinberg, J.B., and Fisher, D.H. Iterate: A conceptual clustering algorithm for data mining. *IEEE Transactions on Systems, Man, and Cybernetics, Part C: Applications and Reviews*, vol. 28, pp. 100–111, 1998.

Fisher, D.H. Knowledge acquisition via incremental conceptual clustering. *Machine Learning*, vol. 2, no. 2, pp. 139–172, 1987. doi:10.1007/BF00114265.

Gennari, J.H., Langley, P.W., and Fisher, D.H. Models of incremental concept formation. *Artificial Intelligence* vol. 40, pp. 11–61, 1989. doi:10.1016/0004-3702(89)90046-5.

Jonyer, I., Cook, D.J., and Holder, L.B. Graph-based hierarchical conceptual clustering. *Journal of Machine Learning Research* vol. 2, pp. 19–43, 2001. doi:10.1162 /153244302760185234.

Talavera, L. and Béjar, J. Generality-based conceptual clustering with probabilistic concepts. *IEEE Transactions on Pattern Analysis and Machine Intelligence* vol. 23, no. 2, pp. 196–206, 2001. doi:10.1109/34.908969.

6

Consumer Co-Creation

Jose Carlos Arce Rioboo

CONTENTS

DEFINITION

Consumer co-creation means fostering individualized interactions and experience outcomes between a consumer and the producers' organization. This can be done throughout the whole product life cycle. Customers may share their needs and comments, and even help spread the word or create communities in the commercialization phase. This approach provides a one-time limited interaction with consumers. Today, it is possible to enable constant interactions to really transfer knowledge, needs, desires, and trends from the consumer in a more structured way: co-creation.

USER

This tool should be used by multidisciplinary teams in which consumers are integrated physically, as members, in a constant dialogue that requires access and transparency to information for both sides, or with a private online community managed through a platform or software.

OFTEN USED IN THE FOLLOWING PHASES OF THE INNOVATIVE PROCESS

The following are the seven phases of the innovative cycle. An X after the phase name indicates that the tool/methodology is used during that specific phase.

- Creation phase X
- Value proposition phase
- Resourcing phase
- Documentation phase
- Production phase X
- Sales/delivery phase
- Performance analysis phase X

TOOL ACTIVITY BY PHASE

- Creation phase—During this phase, customer co-creation is used to ensure that the product or service will be viewed as value added and meet the performance requirements and desires of the people or organizations that receive the output.
- Production phase—During this phase, this tool is used to determine the degree that the product or service provide added value to the customer, and to define problems/improvement opportunities as viewed from the customer standpoint.
- Performance analysis phase—During this phase, this tool is used to determine the degree that the product or service provided added value to the customer.

HOW TO USE THE TOOL

Typically, companies involve consumers using traditional marketing research tools like focus groups. These tools provide a one-time limited interaction with consumers. Today, it is possible to enable constant interactions to really transfer knowledge, needs, desires, and trends from the consumer in a more structured way: co-creation.

Although there are articles before 1999 using the term co-creation, a definition was popularized in the book *The Future of Competition: Co-creating Unique Value with Customers* (Prahalad and Ramaswamy, 2004). The authors defined co-creation as a strategic and intimate joint effort, between firms and consumers, involved in creating value that is unique to the consumer and sustainable to the firm. This definition breaks the traditional concept of a company-centric market and includes high-quality interactions that enable an individual consumer or user to co-create unique products and experiences with the company. These interactions are facilitated by the company.

This definition had been expanded in the book *The Co-creation Paradigm* (Ramaswamy and Ozcan, 2014): "Co-creation is joint creation and evolution of value with stake holding individuals, intensified and enacted through platforms of engagements, virtualized and emergent from ecosystems of capabilities, and actualized and embodied in domains of experiences, expanding wealth–welfare–wellbeing."

Today, the frontiers of the co-creation tasks and methods are fuzzy and include crowdsourcing, user idea contests, tool kits for user innovation or co-design tool kits, communities for user innovation, discussion forums, and the creation of different and creative experience environments (physical and online) in which consumers and users have an active dialogue to co-construct personalized products, services, and business models.

The term *crowdsourcing* was first introduced in 2006 and has many connotations, from simple contributions of many people to help in a task in a voluntary way, collective aggregation of those contributions, reciprocal peer production, to distributed networks of paid labor. These contributions, if creative, can be used in the innovation efforts of an organization. Crowdsourcing has evolved to include crowdfunding.

Different taxonomies have been developed, some of them from autonomous co-creation to whole social product development. One of the main differences between autonomous co-creation and lead users lies first in the motivation; lead users are intrinsically self-motivated to innovate with needs in the edge not met by actual products, without the interactions with companies or manufacturers. In co-creation, organizations provide instruments, tools, and resources to a group of customers or potential customers to create a solution together, even if part of the innovation process is in the side of the customer.

Co-creation means fostering individualized interactions and experience outcomes. This can be done throughout the whole cycle not only of the new development with the generation of ideas and concepts, but also of the product or service life cycle. Customers may share their needs and comments and even help spread the word or create communities in the commercialization phase. The main objective of using co-creation should be to allow optimal customer integration.

There are different ways and philosophies to institutionalize a management process; however, opening the innovation process requires a shift in thinking, a strategy, processes and methodologies, the people, rewards and compensations, culture, etc. First, the company needs to assess whether a particular innovation project is suited for customer co-creation and adjust the portfolio for it.

In the case of co-creation, there is the need to start with building the blocks for the interactions between the firm and the customers that can allow for deep engagement and communication between both parties. It is important to set rules, including the way the intellectual property will be managed and be transparent with the information all the way. Customers need constant feedback, and the information provided by them needs to be analyzed internally.

For having and managing private online communities, Thomas Troch and Tom De Ruyck recommend the following step-by-step approach (Nobel et al., 2014):

1. Define objectives.
2. Select the right technology (which can include forum for discussions, blogs for observations, ideation challenges, survey and poll for numbers, chat for online focus groups, or voice over Internet Protocol for in-depth interviews), specialized features (boards for sketches, simulation software, etc.), profile, and even analytic capabilities.
3. Recruit the right participants.
4. Engage your participants.
5. Set up your interaction guide.

6. Manage the interaction.
7. Analyze the results.
8. Adapt and evolve.

Co-Creation Techniques

Frank Piller and Christoph Ihl had developed a framework to differentiate the basic forms of co-creation (Sigismund et al., 2013). This structure is a 2 × 2 matrix with two dimensions: in the horizontal axis the degree of collaboration (single customer or customer community), and in the vertical axis the degree of freedom (low and high).

The degree of collaboration refers to "the structure of the underlying relationships in an open innovation setting." The degree of freedom refers to "the nature of the task that has been assigned to customers." The framework yields four kinds of co-creation methods:

- For a single customer with a low degree of freedom with predefined and narrow task: co-design tool kits
- For a single customer with a high degree of freedom with creative and open task: idea contests
- For a customer community or network and a low degree of freedom: discussion forums
- For a customer community or network and high degree of freedom: social product development

Idea Contests or Innovation Tournaments

An innovation tournament has been defined by Terwiesch and Ulrich (2009) as a competition among opportunities that includes multiple rounds of competition.

It begins with a request asking for solutions to a given problem, feature, task, technology, or expected solution. There can be one-time contests or open permanently. It is important to state the problem carefully.

There is a reward offering for the winning solution, and it is crucial to incentivize participants to transfer their ideas to a company. These rewards could be cash, licensing contracts, nonmonetary acknowledgments (recognitions), etc. With it, a time frame has to be set for responses. Hopefully, a large set of ideas, opportunities, or concepts will be proposed. A filtering process selects subsets that move to subsequent rounds, until one or more are picked as champions for full development. Thus, an idea screening or evaluation process must be set before the contest and clear to all participants. The way that the ideas could be screened varies a lot: from customer voting, panels of experts with a set of evaluation criteria, and even prediction and virtual stock markets.

In a co-creation effort, the tournament or competition is open to the public or to selected customers or communities of them. In open innovation, it could include intermediary markets or companies like InnoCentive, Nine Sigma, yet2com, etc.

Co-Design Tools

The company provides a tool kit that enables customers to create their own designs, solving their particular needs, desires, or tastes; or create a concept, by trying out and experimenting with different combinations of chemicals, platforms, modules, components, parameters, colors, flavors, or basic designs. This is possible in the boundaries set by the tool kit or simulation, drawing, or design software provided.

Discussion Forums

In maintaining a discussion forum, the company needs to have the correct infrastructure and human resources to organize and facilitate a constant and interesting dialogue with focus on value creation, for both parties. The motivation and self-election of customers to participate are different than in the other two tools. The discussion forums need to provide a platform for users to exchange experiences, data, tips, and support for each other in different formats (pictures, videos, messaging, etc.). The employees involved must have the expertise to guide in good terms those discussions and avoid problems that could damage the brand.

Social Product Development

They are also called communities in creation. Different from crowdsourcing, in a private community for co-creation, a smaller group of people (well chosen before or after an open invitation) are engaged in new product or service development. The main goal is to go in-depth on its development with a rich interchange of ideas, issues, and understandings, by being connected in an intense way for a short period of time in a private dedicated platform.

In a more social open community and into the crowdsourcing spectrum, the company engages consumers in many different activities along the entire span of the innovation process, which can include financing.

EXAMPLES

The most used but representative examples for co-creation are as follows:

- *Threadless* (http://www.threadless.com): A fashion T-shirt company, where consumers design their own colorful T-shirts. Consumers approve by consensus before any investment is made in a new product. Once the minimum number of customers who express their explicit willingness to buy a particular design is reached, that design goes to production. The selected designers receive a prize in cash.
- *LEGO* (http://www.lego.cuusoo.com): A company that helps you design a product to your own specifications.
- *Quirky* (http://www.quirky.com): A manufacturer and seller of customer-designed products. Individuals suggest new concepts of products and the community votes on the best's ideas while committing to purchase that product.

SOFTWARE

Almost all major innovation management commercial platforms support co-creation programs. Just to mention some of them:

- Hype Innovation: http://www.hype.com
- Brightidea: http://www.brightidea.com
- Spigit: http://www.spigit.com
- Imaginatik: http://www.imaginatik.com
- InnovationCast: http://www.innovationcast.com
- Cognistreamer: http://www.cognistreamer.com
- Inno360: http://www.inno-360.com
- Qmarkets: http://www.qmarkets.com
- InnoCentive@Work: http://www.innocentive.com/innovation-solutions/innocentive-at-work

REFERENCES

Nobel, C.H., Durmusoglu, S.S., and Griffin, A. *Open Innovation. New Product Development Essentials from the PDMA*. Chapter 6, pp. 135–172. Hoboken, NJ: John Wiley & Sons Inc., 2014.

Prahalad, C.K. and Ramaswamy, V. *The Future of Competition: Co-creating Value with Customers*. Boston: Harvard Business School Press, 2004.

Ramaswamy, V. and Ozcan, K. *The Co-Creation Paradigm*. Stanford, CA: Stanford University Press, 2014.

Sigismund Hoff, A., Moslein, K.M., and Reichwald, R., eds. *Leading Open Innovation*. Chapter 9, pp. 139–154. Cambridge, MA: MIT Press, 2013.

Terwiesch, C. and Ulrich, K.T. *Innovation Tournaments: Creating and Selecting Exceptional Opportunities*. Boston: Harvard Business Press, 2009.

SUGGESTED ADDITIONAL READING

Pilkler, F.T. and Ihl, C. *Open Innovation with Customers: Foundations, Competences and International Trends*. Expert study commissioned by the European Union, The German Federal Ministry of Research, and the European Social Fund, published as part of the project "International Monitoring," Aachen, Germany, 2009.

7

Creative Thinking

Frank Voehl

CONTENTS

DEFINITION

Creative thinking is all about finding fresh and innovative solutions to problems, and identifying opportunities to improve the way that we do things, along with finding and developing new and different ideas. It can be described as a way of looking at problems or situations from a fresh perspective that suggests unorthodox solutions, which may look unsettling at first.

USER

This tool can be used by individuals or groups, but its best use is with a group of four to eight people. Cross-functional teams usually yield the best results from this activity.

OFTEN USED IN THE FOLLOWING PHASES OF THE INNOVATIVE PROCESS

The following are the seven phases of the innovative cycle. An X after the phase name indicates that the tool/methodology is used during that specific phase.

- Creation phase X
- Value proposition phase X
- Resourcing phase X
- Documentation phase X
- Production phase X
- Sales/delivery phase X
- Performance analysis phase X

TOOL ACTIVITY BY PHASE

Although creative thinking is normally considered as being part of only the creative, production, and sales and delivery phases, it actually applies to all the phases. There is a strong need for new and unique approaches to the way we approach preparing value propositions, financing our projects, documenting our activities, and analyzing how well the projects are. In an innovative organization, innovation and creativity apply to all parts of the organization.

HOW TO USE THE TOOL

Creative thinking is all about finding fresh and innovative solutions to problems, and identifying opportunities to improve the way that we do things, along with finding and developing new and different ideas. It can be described as a way of looking at problems or situations from a fresh perspective that suggests unorthodox solutions, which may look unsettling at first.

- *Creativity* is developing new or different ideas.
- *Innovation* is converting ideas into tangible products, services, or processes. Creativity without innovation is wasted effort. The challenge that every organization faces is how to convert good ideas into profit. That is what the creativity process is all about.

This means that the organization has to have a system in place that will put an individual's ideas on a fast track through the process of getting it approved and implemented. This requires that the creative/innovative process make a smooth transition from an individual, to a team, to all of the affected individuals.

Creative thinking can be stimulated both by an unstructured process such as brainstorming or brainwriting, and by structured methods or processes such as lateral thinking.

There are many tools for creative thinking in the innovation literature, as popularized in the following seven major works on the subject:

- Edward de Bono presents 13 tools in his book *Serious Creativity*.
- Grace McGartland has 25 tips and techniques in *Thunderbolt Thinking*.
- Arthur VanGundy covers 29 tools in *Idea Power*.
- Michael Michalko describes 34 techniques in *Thinkertoys*.
- Roger von Oech has 64 methods in his *Creative Whack Pack*.
- Koberg and Bagnall give guidance on 67 tools in *The Universal Traveler*.
- James Higgins tops them all with his book *101 Creative Problem Solving Techniques*.

While there is an overlap among these popular works, there are at least 250 unique tools in these seven books alone. And these are only a few of the references available on the topic of creative thinking, as shown in the associated Suggested Additional Reading at the end of this chapter.

Start by taking the MindTools Quickscore 3-Minute Creativity Test, which helps you assess and develop your business creativity skills (http://www.mindtools.com/community/pages/article/creativity-quiz.php).

We must emphasize that creativity is all about finding fresh and innovative solutions to problems, and identifying opportunities to improve the way that we do things, along with finding and developing new and different ideas. As such, any one of us can be creative, as long as we have the right mind-set and use the right tools. This test helps you to think about how creative you are right now. Take it, and then use the tools and discussions that follow to bring intense creativity to your everyday work.

Organizing the Organization for Creative Thinking

The creative thinking methodology uses the following process for accomplishing this type of organization:

1. Embed creativity into the organization's culture and vision.
2. Assess creativity status. An assessment of the creativity performance level of the organization should be made. Typical questions that would be answered are
   - Does the organization have a measurement of its return on its creativity investment?

- What percentage of our effort is devoted to creative activities, and is that enough?
- Does the organization have a chief creativity officer?
- Do we have creativity goals and targets?
- What percentage of our employees made a measurable creativity improvement in the past 12 months, and is that percentage high enough?
- Are resources made available to support the refinement of new ideas?
- What roadblocks are in the way of the organization becoming more creative?

3. Establish a creative thought process (see Figure 7.1).
4. Train everyone in how to be creative.
5. Set up an idea review system that will quickly bring to upper management's attention the ideas that will have an important impact on the organization's present or future performance so that these ideas can be implemented quickly. Lower-level review boards should be established to expedite the evaluation of less important ideas and assume the responsibility for implementing them.
6. Budgets should be set up for the review board to fund creative proposals. The review board's return on investment related to its budget should be measured. At a very minimum, the return on investment should be 12:1 if the process is working effectively.
7. Management should recognize and support employees who have an entrepreneurial attitude by having them form *possibility teams* to explore, develop, and prepare business proposals for good concepts.
8. Reward systems that reward both noble failures and crowning successes should be established to reinforce a risk-taking environment throughout the organization.

The tools of creative thinking are simply various combinations of practical ways to implement this heuristic eight-step process—*to focus attention, to escape the current reality*, and *to continue mental movement*. The relative weights given to attention, escape, and movement, and the mechanics of directing these three mental actions, vary among the methods. But this variation makes sense because each situation we encounter is different, each group is different, and each person is different. Once we understand these three basic principles, we can adapt techniques to suit various needs, situations, and personalities.

Focus Principle

Creativity requires that we first focus our attention on something (to what?), typically something that we have not focused much attention on before. The primary innovation of the Apple Macintosh computer in the early 1980s was that its designers focused not on raw computing power but on the user interface. By focusing attention on things that are normally taken for granted (in this case, the command line interface predominant in the early 1980s), creative thinking techniques prepare our minds for breakthroughs (here, the graphical user interface). Ask the question: To what? ...

- Elements in the current reality
- Features, attributes, and categories
- Assumptions, patterns, and paradigms
- Metaphors and analogies
- What works and does not work
- Anything you do not normally pay attention to

Escape Principle

Having focused our attention on the way things are currently done, the second principle behind all creative thinking methods calls us to mentally escape our current patterns of thinking. For example, Edward de Bono suggests that we use the *po* tool to signal our intention to make a statement of mental escape. To a group working to decrease the time that customers wait to receive a service, we might say, "Po, they have passed a law making it illegal for customers to wait more than 30 seconds; what are we going to do now?" The statement invites us to escape our current paradigm about customer flow and, for a moment, imagine a very different world.

The principle of escape explains why a simple walk in the woods can bring about creative thoughts. When we walk in the woods, we escape the confines of the current ways, both mentally and physically. Similarly, staring at yourself in the mirror while you shave or put on make-up provides a momentary mental escape that may allow a novel mental connection about a work problem to emerge. Ask the question: From what? ...

- Current mental patterns
- Time and place
- Early judgment
- Barriers and rules
- Your past experiences

Movement Principle

Simply paying attention to something and escaping current thinking on it is not always sufficient to generate creative ideas. Unfortunately, the natural mental processes of judgment tend to reject new thoughts as not productive or too ridiculous to dwell on. Movement—the third underlying principle behind the diverse tools of creative thinking—calls us to keep exploring and connecting our thoughts.

Movement is a key principle behind the classic creative thinking technique of brainstorming. The ground rules of brainstorming are to generate as many ideas as you can, with no criticism, building on the ideas of others. In other words, keep moving. Similarly, asking a group to come up with a sketch that illustrates their vision of the company's future is also a movement technique. You cannot simply state the vision and be done with it; your mind must dwell on it long enough to complete the sketch. During that time, the mind—which is never idle—generates new connections and ideas that might expand the basic concept. Ask the question: In what sense? ...

- In time or place
- To another point of view
- Free association
- Building on ideas

ADVANTAGES

Every new product, program, or service results from an idea that then follows an innovation cycle of testing, implementing, and marketing. It is generally accepted that the key to competitive advantage is generating and successfully exploiting new ideas that come from creative thinking of both individuals and groups.

DISADVANTAGES

The environment in some organizations may prove to be hostile to creative thinking. Anyone managing this creative thinking process is likely to encounter obstacles such as

- General resistance to change
- Free expression being stifled by a pervasive culture of blame
- Failure being regarded as a cause of penalties and not an opportunity to learn
- Inadequate and nonexistent incentives leading to slow decision making
- Reluctance to think and move outside of strict job descriptions
- A view that the best ideas come from the top
- Rigid formalities and rules

SUMMARY

One factor that strongly affects an organization's creativity success rate is its attitude toward creativity and problem–opportunity finding in the first place. On the creativity spectrum, it ranges from inactive, active, and proactive to hyperactive. Regardless of the type of approach, creative people do not just sit around and wait for opportunities or problems to surface. Instead, they scan their environment for potential opportunities or issues, and they see this as exercising creativity time well spent, for in actuality they are often excited by the opportunity to change things. They are not intimidated by change; rather, they embrace it.

The tools of creative thinking were combined with various combinations of practical critical thinking ways to develop procedural knowledge and implement the creative thinking process while obtaining feedback from the environment—*to focus attention, to escape the current reality,* and *to continue mental movement.*

Focus Attention

1. Creative thinking is a cognitive activity that may result in a creative production that groups or individuals perceive as useful and new.

The products may be pieces of writing such as books, essays, poems, or short stories; physical creations such as new robots, works of art, buildings, or miniature representations; new systems, theories, or conceptualizations such as quality circles, managements by objective, the wave theory of light, the self-concept theory in psychology, or the periodic theory of elements in chemistry; performances in drama, music, dance, or speech; or inventions such as automobiles, the airplane, or the automatic can opener. We call the products creative if they represent a transformation or a reconceptualization, have aesthetic coherence and appeal, represent a new configuration or connection of ideas, or serve some functional or explanatory purpose. Problem solutions have all these critical elements, plus relevance or resolution to the original problem (Isaksen et al., 1993).

2. Creative intelligence is involved when skills are used to create, invent, discover, imagine, suppose, or hypothesize. Creativity is one of three sets of abilities (the other are practical and analytical) that are integrated "to attain success in life, however an individual defines it, within his or her sociocultural context" (Sternberg and Grigorenko, 1997).

3. Creative thinking involves the realization of an analogy between previously unassociated mental elements.

4. Creative thinking is among the most complex of human behaviors. It seems to be influenced by a wide array of developmental, social, and educational experiences, and it manifests itself in different ways in a variety of domains. The highest achievements in the arts are characterized by their creativity, as are those in the sciences. Creativity is also quite common in a wide range of everyday activities. Theories of creativity have attempted to recognize the inherent complexity by defining creativity as a syndrome, or even a complex (Runco and Sakamoto in Sternberg, 1999).

Escape Current Reality

5. Creative thinking is a novel and useful idea or product for creating a new reality.

6. Creativity is the confluence of intrinsic motivation, domain-relevant knowledge and abilities, and creativity-relevant skills; the latter includes coping with complexities, knowledge of problem-solving heuristics, concentration, ability to set aside problems, and high energy (Sternberg and Grigorenko, 1997).

7. Creativity is the result of an anomaly with a system, or moderate asynchronies between the individual, domain, and field (Gardner, 2011).

Continue Mental Movement

8. Creativity is produced by a confluence of six distinct but interrelated resources: intellectual abilities, knowledge, styles of thinking, personality, motivation, and environment (Sternberg and Lubart in Sternberg, 1999).
9. Creativity is a creative product produced by a creative person engaged in a creative process within a creative environment.
10. Creativity is an essential life skill through which people can develop their potential to use their imagination to express themselves, and make original and valued choices in their lives.
11. Creativity is the exploration and transformation of conceptual spaces (Margaret Boden). (A conceptual space is smaller than a domain.)
12. Creativity arises out of the tension between the rules and imagination (Ian Hodder).

EXAMPLE

A Team Development Session for Understanding Creative Thinking at the South Wales Mental Health Advocacy Center Using Critical Thinking Models

Objective: To provide the staff group with an understanding of creative thinking that they would be able to use in their day-to-day work.

Process: With just two hours to address the objective, the session began with an overview of the creative process using the Example Critical Thinking Model (see Figure 7.1). Participants then warmed up for later activities by establishing a new belief by using drawing exercises and Play-Doh to create visualization and knowledge images. After a discussion around areas for development, a single key challenge was identified and the model making began. The group was

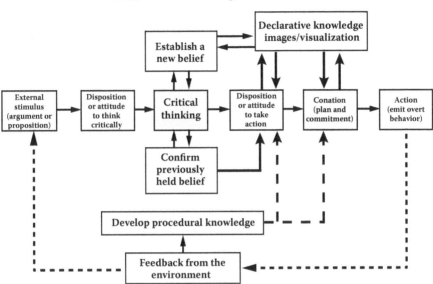

Model of critical thinking and its modification

FIGURE 7.1
Critical thinking model.

split into three subgroups, each given a model to create a relation to confirm previously held beliefs, their view of the organization, their role, and a perspective from the person they worked with. The tools of creative thinking were combined with various combinations of practical critical thinking ways to develop procedural knowledge and implement the creative thinking process while obtaining feedback from the environment—*to focus attention, to escape the current reality,* and *to continue mental movement.*

Result: From the laughter in the room, it was evident that people obviously had fun undertaking the exercises and feeding back to each other. They also learned something about creativity and skills they could use with the people they worked with.

Comments from the team:

We are usually negative about these exercises and their impact, but certain approaches/ideas/actions struck home; Will definitely use some of these techniques at our peer support meeting; Very enjoyable and informative. Thank you; Good stuff, enjoyed it.

CASE STUDY

The Teagle Creative Thinking Working Group of the Five Colleges of Ohio

This case study involved the investigation of the following critical thinking issues:

- Definitions of creativity and critical thinking
- The students' and colleagues' understandings of the nature and function of creativity in the liberal arts setting
- Processes and products within the liberal arts setting that stimulate or hinder creativity
- Explanations for how creative and creative/critical thinking, the latter in practice long the sentry of liberal arts, complement and support each other
- Methods to evaluate the liberal arts setting as a venue for nurturing creativity across the curriculum

Stage One of the case study project focused on various definitions of creative thinking, identification of human qualities, and other factors associated with creativity. It involved the construction of indirect and direct measures to evaluate creative thinking within the classroom, and personal narratives about how their Stage One work has affected teaching, attitude toward students, and their scholarly pursuits.

Theoretical Context for the Case Study

Enhancing one's creative abilities has long been a tacit assumption underlying liberal education in the Western world, but little research exists to help students and educators determine exactly what is meant by creative thinking in the classroom, let alone methods by which we can promote and evaluate the success of such thinking. Some even hypothesized that general and field-specific undergraduate education ironically works most effectively to squash creativity. Yet there remains something in all post-secondary teachers that recognizes the reality of creative thinking in research and teaching, and values that same reality for students.

The field of psychology has only embraced the serious study of creativity since the mid-20th century, and much of the literature prior to that date situated the study of creative thinking as the provenance of philosophy, mysticism, and spirituality. As a result, creative thinking research has been beset with the belief that the source of creativity resides in the divine, the *mad*, the specially gifted—that *Ah-Hah* or *Eureka* moment generated from the unknowable or bizarre.

Within the last 50 years, however, psychology has focused creativity research on systems theories that identify stages of unilinear development and on readily identifiable criteria or variables of creativity across and within specific domains. While still hampered by problems defining creativity and creative thinking as well as narrow visions of creativity that inappropriately generalize human behavior outside the sociocultural context (Sternberg, 1999), this redirection has foregrounded cognitive, affective, and environmental factors that demystify creativity. The case study postulated that psychometrics had not yet provided a reliable standardized test for creativity—and perhaps never will, but found that creativity was no longer the sole property of the Einsteins, Van Goghs, and Dickinsons of the world—it is a quality inherent in greater or less degrees in all humans, one necessary for a successful life.

Consequently, educators now have a small yet powerful body of literature with which to move creativity studies onto the campus and into the classroom—to determine how to amplify an individual's creativity to position our students for the dynamics of the 21st century.

SOFTWARE

- *The Creative Thinker by Idon Resources.* If you are looking for new perspectives, wish to develop your imagination, make better decisions, and take more effective action, then welcome to the world of the Creative Thinker. This new software brings you nonlinear, yet structured, thinking with the ability to further develop your thoughts in the form of user-friendly and fun-to-use graphical hexagon modeling. You can even organize your knowledge, supplement ideas with notes, and directly and meaningfully link ideas to the Internet, documents, slide presentations, spreadsheets, and more.

The Creative Thinker literally allows you to visualize your thoughts, rapidly access knowledge, and combine the two creatively in real time for insights. See http://www.idonresources.com/ct/creative thinker.html.

- *Creative Thinking Program.* This program can be customized to be anywhere from two days to six months. The objective is to bring innovation, improvisation, and creative thinking into a team or organization. It consists of a combination of skill development training, real-time process facilitation, team and individual coaching, and interviews. This program includes a repeatable creative thinking process that can be used by the different teams, along with relative thinking best practices, tools, and techniques for individuals and teams. See http://www.creativeemergence.com/creativethinking.html.

- *MindJet. The Platform for Enterprise Innovation.* More than collaborative, more than social, MindJet is the engine for innovation programs that keep the crowd engaged. See http://www.mindjet.com/.

- *Insight Maker.* Insight Maker runs in your web browser. No software download or plug-ins are needed. Get started building your rich pictures, simulation models, and insights. For further information, see http://insightmaker.com/.

- *X-Mind.* You can use simple mind maps if you choose, or *fishbone*-style flowcharts if you prefer. You can even add images and icons to differentiate parts of a project or specific ideas, add links and multimedia to each item, and more. See http://www.xmind.net/.

REFERENCES

de Bono, E. *Serious Creativity.* New York: HarperCollins Publishing, 1992.

Gardner, H. *The Theory of Multiple Intelligences: As Psychology, As Education, As Social Science.* Available at https://howardgardner01.files.wordpress.com/2012/06/473 -madrid-oct-22-2011.pdf.

Higgins, J.M. *101 Creative Problem Solving Techniques.* Winter Park, FL: New Management Publishing Company, 1994.

Isaksen, C., Puccio, G., and Treffinger, D. An ecological approach to creativity research: Profiling for creative problem solving. *Journal of Creative Behavior*, 27(3), 149–170, September 1993.

Koberg, D. and Bagnall, J. *The All New Universal Traveler: A Soft-Systems Guide to Creativity, Problem-Solving, and the Process Of Reaching Goals.* Los Altos, CA: William Kaufmann Inc., 1981.

McGartland, G. *Thunderbolt Thinking: Transform Your Insights and Options into Powerful Results*. Austin, TX: Bernard-Davis, 1994.

Michalko, M. *Thinkertoys: A Handbook of Business Creativity for the 90s*. Berkeley, CA: Ten Speed Press, 1991.

Sloane, P. *The Leader's Guide to Lateral Thinking Skills: Unlocking the Creative Thinking Within*. London: Kogan Page, 2004. ISBN 0749447974.

Sternberg, R. and Grigorenko, E., Are Cognitive Styles Still in Style, Yale University, New Haven, CN: American Psychologist, vol. 52, no. 7, July 1997.

VanGundy, A.B. *Idea Power*. New York: American Management Association, 1992.

von Oech, R. *A Whack on the Side of the Head*. New York: Warner Books, 1983.

SUGGESTED ADDITIONAL READING

Brunner, G. *Primary Analogies: Critical and Creative Thinking*. Cambridge, MA: Educators Pub. Service. ISBN 0838822843.

Copeland, M. *Socratic Circles: Fostering Critical and Creative Thinking*. Portland, ME: Stenhouse Publishers. ISBN 1571103945.

Ebert, C. and Ebert, E.S. *The Inventive Mind in Science: Creative Thinking Activities*. Hillsdale, NJ: Teacher Ideas Press.

Hussey, W. *Brilliant Activities to Stimulate Creative Thinking* (contains over 150 entertaining, open-ended challenges providing mental stimulation for all). Bedfordshire, UK: Brilliant Publications. ISBN 1783170212.

Inch, E.S. and Warnick, B. *Critical Thinking and Communication: The Use of Reason in Argument*. Boston: Allyn & Bacon. ISBN 0205672930.

Libonate, G. and Brunner, G. *Six Thinking Hats De Bono, Edward Self-Help, Creativity Education Decisions*. Cambridge, MA: Educators Publishing Service. ISBN 0838822908.

Marcus, S. *World Kids Parents' Guide to Creative Thinking Class Set*. San Antonio, TX: Susie Monday Foundry Media, 2009. ISBN 061519060X.

Monahan, T. *The Do-It-Yourself Lobotomy: Open Your Mind to Greater Creative Thinking*. New York: Wiley. ISBN 0471417424.

Nadler, G. and Hibino, S. *Breakthrough Thinking*, 2nd Ed. Roklin, CA: Prima, 1994.

Pesut, D.J. and Herman, J. *Clinical Reasoning: The Art and Science of Critical and Creative Thinking*. Delmar. ISBN 0827378696.

Plsek, P.E. *Creativity, Innovation, and Quality*. Milwaukee, WI: ASQC Quality Press, 1997.

Richard, E. *Critical (Creative) Thinking Card Games: Easy-to-Play, Reproducible Cards*. Scholastic Teaching Resources.

Ritter, D. and Brassard, M. *The Creativity Tools Memory Jogger: A Pocket Guide for Creative Thinking*. GOAL/QPC, 1998. ISBN 1576810216.

Ruggiero, V. *The Art of Thinking: A Guide to Critical and Creative Thought*. New York: Pearson Longman, 2011. ISBN 03214112.

Shingō, S. *Kaizen and the Art of Creative Thinking: The Scientific Thinking Mechanism*. New York: Enna Products Corporation, 2007. ISBN 1897363591.

Vogel, T. *Breakthrough Thinking: A Guide to Creative Thinking and Idea Generation*. Blue Ash, Ohio: F & W Media Incorporated, 2014.

Wonder, J. and Donovan, P. *Whole Brain Thinking*. New York: Ballantine, 1984.

8

Directed/Focused/ Structured Innovation

Maria B. Thompson

CONTENTS

To raise new questions, new possibilities, to regard old problems from a new angle requires creative imagination and marks real advances in science.

Albert Einstein

DEFINITION

Directed Innovation is a systematic approach that helps cross-functional teams apply problem-solving methods like brainstorming, TRIZ (theory of inventive problem solving), creative problem solving, Six Thinking Hats™, Lateral Thinking™, Assumption Storming™, inventing, Question Banking™, and Provocation to a well-defined problem in order to create novel and patentable solutions.

USER

This tool is useful for individual creative problem solving, but is best applied with an agnostic facilitator, a complex or difficult problem space, and a diverse group ranging from 8 to 25 in size working in pairs. Cross-functional teams, where a marketing or product manager is paired with an engineer or field service technician, usually yield the best number and quality of solutions.

OFTEN USED IN THE FOLLOWING PHASES OF THE INNOVATION PROCESS

The following are the seven phases of the innovation cycle. An X after the phase name indicates that the tool/methodology is used during that specific phase.

- Creation phase X
- Value proposition phase X
- Resourcing phase
- Documentation phase X

- Production phase
- Sales/delivery phase
- Performance analysis phase

TOOL ACTIVITY BY PHASE

- Creation phase—Directed Innovation is most frequently used in the creation phase of innovation processes in order to (a) explore the problem domain in detail, surfacing any assumptions about the solution space; (b) generate a generic, clearly-defined set of open-ended, thought-provoking questions for a diverse team to ideate with; (c) brainstorm many alternative solutions to a gnarly or complex problem in order to select a novel, feasible solution that will delight the customer and leverage the competencies of the inventors; and (d) ensure there is white space to play in, and no other entities have already locked up intellectual property rights to prevent others from introducing new products or services into the market or to the same end users.
- Value proposition phase—The value proposition can be derived from evaluating the alternative solutions generated during ideation according to novelty (e.g., patentability or trade secret worthiness), and impact or importance to the customer's or end user's operations (e.g., *jobs to be done*).
- Documentation phase—In the documentation phase of the innovation process, the team captures patent disclosures and provisional patent applications (or appropriate intellectual property documentation) to protect the novel aspects of the chosen solution. Design documentation is written by building a requirements matrix based on the problems and questions addressed by the alternative solutions generated in ideation and mapping these to the selected solution's design parameters, and testing that these perform as expected in the final end product or service.

HOW TO USE THE TOOL

Introduction

The following is the nine-step approach to conducting a Directed Innovation activity (see Figure 8.1).

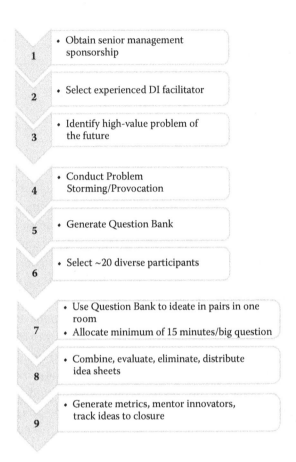

1 • Obtain senior management sponsorship

2 • Select experienced DI facilitator

3 • Identify high-value problem of the future

4 • Conduct Problem Storming/Provocation

5 • Generate Question Bank

6 • Select ~20 diverse participants

7 • Use Question Bank to ideate in pairs in one room
  • Allocate minimum of 15 minutes/big question

8 • Combine, evaluate, eliminate, distribute idea sheets

9 • Generate metrics, mentor innovators, track ideas to closure

FIGURE 8.1
Directed Innovation (DI) process.

Step 1: Obtain Senior Management Sponsorship

Directed Innovation should only be used when a small team or an individual is not able to solve a complex high-value problem on their own using other well-known creativity tools as discussed in this book. Senior management sponsorship is critical. The problem domain should be of strategic importance to the growth of the company. If so, a senior manager should be willing to invest the time of their most valuable people and budget for requisite follow-up on the selected idea(s), including prototyping, commercialization, and patent or trademark filings. You do not want to go to the expense and trouble of planning the ideation session and conducting it only to discover that no one in management is willing to assign development resources. The next time you try to host an ideation session, it will be difficult, if not impossible, to convince the creative thinkers to participate. Do not move forward to subsequent steps until you are certain that the investment in Directed Innovation is both warranted and supported.

Step 2: Select an Experienced Agnostic Facilitator

It is important to have a trained, seasoned facilitator who knows how to get the mindshare of all participants in both the preparation and planning phase, as well as the ideation and solution evaluation phases. The facilitator should be agnostic with respect to the problem or technology domain. It is a mistake to have someone who is a subject matter expert (SME) plan or facilitate the ideation session. They are typically too close to the subject matter, and know too much about *what will and will not work*, *what has been tried before and failed*, and are often too attached to the outcome or invested in the result, which might include their own pet solution or patent filings. Often the best solutions come from those with no subject matter expertise in the problem domain.

You might think that it would be good to have a patent attorney facilitate or scribe the ideation portion of the meeting, though this is inadvisable. With years of experience facilitating these sessions, we have learned that whoever holds the pen, holds the power. In the past, we had patent attorneys lead the meetings with easels and try to scribe all ideas as participants shouted them out. This had three major issues: (a) the introverts did not call out their ideas, (b) the overall group did not generate as many ideas or creative ideas due to the *group think* influence of the extroverts and SMEs, and (c) the attorney scribes subconsciously filtered their note-taking in real-time to capture only those ideas they felt were easily captured in a patent claim.

Step 3: Identify a High-Value Problem of the Future

Again, Directed Innovation is not for simple problems. Figure 8.2 shows the five levels of invention complexity. Level 1 is the lowest level of complexity and level 5 is the true creation of a new concept often referred to as the *Eureka level*. The complexity of levels 1 through 3 is such that it does not require the use of Directed Innovation, and is often solved by simpler approaches like brainstorming. Directed Innovation is a creative problem-solving method that should be used on complex, high-value problems that are critical to the future growth strategy of the company. Typically, the

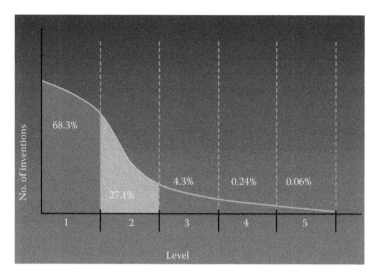

FIGURE 8.2
Five levels of innovation complexity.

lead technologists, or senior managers, are the originators of the topic or problem domain for Directed Innovation sessions. The high-value aspect can be assessed by determining whether you, your competitors, and your customers have been trying to solve the problem, with inadequate or incomplete solutions, for a relatively significant period of time. Often, this can be determined not only by searching for publications about the problem from industry players on the Internet, but from capturing the language (aka, jargon) of the problem and searching the worldwide patent portfolio "Background of the Invention" section of published patents and applications to determine whether there are several older or recent filings or pending applications. This search and exposure to prior art should be limited to one to two SMEs involved in the up-front planning of the ideation session, so as not to anchor the ideation participants to the existing prior art. At this point in the Directed Innovation process, critical thinkers are highly valuable, since they will know all the problems and obstacles that have thwarted successful solution generation in the past. They can highlight the inadequacies of existing solutions, whether our solutions or our competitors' solutions.

Step 4: Conduct Problem Storming/Provocation

Problem Storming and Provocation originated from a design session conducted with Motorola in the late 1990s at an internal innovation conference. Provocation resembles Edward de Bono's Lateral Thinking™ method, though it has been customized for the purposes of Directed Innovation with additional Assumption Storming to enable a deep exploration of the original problem from many different angles (see Figure 8.3).

Figure 8.3 shows a worksheet to capture the output of Problem Storming, Assumption Storming, and Provocation exercises. There can be multiple Provocation worksheets generated for each gnarly problem or problem domain.

The first step is for the original problem to be recorded as a goal or objective in #1 (the middle box or Insight #1) in Figure 8.3. It works best if the problem can be stated as a "How might we ... overcome xxx obstacle?" or "What are all the ways we might overcome/eliminate ... problem/parameter/function?" question.

In step #2 in Figure 8.3, the arms diagonally off the #1 Problem Statement in Figure 8.3, the SMEs work with the facilitator to generate all the discrete constraints, limitations, and assumptions that currently prevent an ideal solution from being generated. Often, these are stated as design constraints, dogma, and known requirements bounding the scope of the solution space. A way to facilitate comprehensive generation of these is to look for all the statements such as "we would do xxx, BUT..." or "we have tried yyy, BUT...." Assumption Storming (see Figure 8.4) is then used to enable the SMEs to perform a comprehensive check to ensure they have surfaced all the constraints, limitations, and beliefs that must be considered before exploring potential opportunities. It is important that each of the constraints generated in #2 of Figure 8.3 is a discrete constraint and not a combination of multiple design characteristics or parameters describing the attributes of existing problematic solutions or the anticipated ideal solution. The advice to SMEs should be to list all the design requirements or parameters that must exist for the solution to be viable. These items are all adjectives, adjective phrases, and descriptors. Several Provocation worksheets may be necessary, depending on the breadth and complexity of the original problem statement or objective in #1 in Figure 8.3.

In #3 in Figure 8.3, the facilitator walks the SMEs one by one through each individual constraint and questions WHY this constraint is real, necessary, and needs to be taken into account before an adequate solution can be generated. You are listing the underlying rationale or reason for the design constraint in #2, one at a time. Also, each time you list one reason WHY the constraint in #2 is important, you ask yourself again

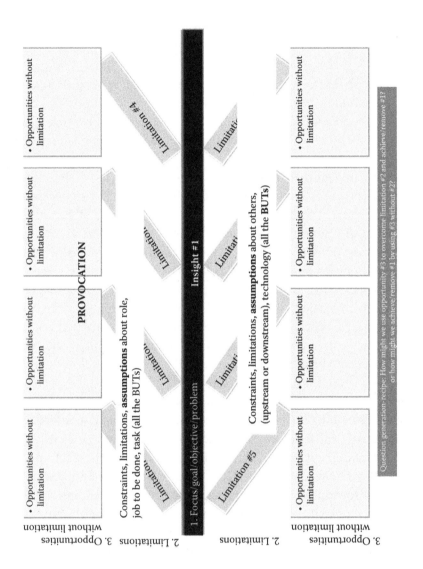

FIGURE 8.3

Provocation worksheet for ideation prework.

IGNORE OR BUST CONSTRAINTS
with TRIZ
*Question*Bank™

99 Questions based on 40 TRIZ Principles - v1

1. Segmentation (Principle #1)
1. How might it be segmented?
2. How might it be segmented into independent parts?
3. How might it be easy to disassemble?
4. How might we increase the degree of fragmentation or segmentation?

2. Separation (Principle #2)
5. How might the interfering parts or properties be singled out?
6. How might only the necessary part be single out?

3. Local Quality (Principle #3)
7. How might the structure be changed from uniform to non-uniform?
8. How might the external environment or influence be changed from uniform to non uniform?
9. How might each part function in conditions most suitable for its operation?
10. How might each part fulfill different and useful functions?

4. Symmetry Change (Principle #4)
11. How might the shape be changed from symmetrical to asymmetrical?
12. If it is asymmetrical, how might the degree of asymmetry be increased?

5. Merging (Principle #5)
13. How might identical or similar objects be brought closer together or merged?
14. How might identical or similar parts be assembled to perform parallel operations?
15. How might operations be contiguous or parallel?
16. How might operations be brought together in time?

6. Multifunctionality (Principle #6)
17. How might parts or objects perform multiple functions?
18. How might parts or objects eliminate the need for other parts?

7. Nested Doll (Principle #7)
19. How might one object be placed inside another?
20. How might one object be placed inside another, and then inside another?
21. How might one part pass through a cavity into another?

8. Weight Compensation (Principle #8)
22. How might the weight of an object be compensated by merging with other objects to provide lift?
23. How might the weight of an object be compensated by interacting with the environment?
24. How might the weight of an object be compensated by interacting with the aerodynamic forces?
25. How might the weight of an object be compensated by interacting with the hydrodynamic forces?
26. How might the weight of an object be compensated by interacting with the aerodynamic buoyant forces?

FIGURE 8.4
TRIZ Question Bank of inventive principles for assumption busting.

WHY the first reason listed in #3 is important. You ask yourself WHY three to five times until you are sure you have the root reason WHY the original constraint in #2 is still valid and important to consider. Often, the SMEs realize that the constraint listed in #2 is an invalid assumption and not a limitation on the solution. Creative SMEs will even generate creative solutions at this stage, or at least new opportunities that can be explored in the later ideation session. The most experienced SMEs have used TRIZ (described in Chapter 29 of this book) to take each constraint in #2 of Figure 8.3 and treat it like a contradiction or trade-off to be overcome.

Assumption Storming Question Bank for Provocation

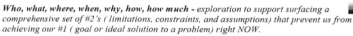

Assumption Storming
Question Bank

Who, what, where, when, why, how, how much - exploration to support surfacing a comprehensive set of #2's (limitations, constraints, and assumptions) that prevent us from achieving our #1 (goal or ideal solution to a problem) right NOW.

Categories the facilitator uses with 2-3 SMEs from Observation domain to elicit #2's in Provocation (planning DI Ideation sessions)

I.   WHO are all the players, stakeholders, and gatekeepers we might consider that influence our ability to achieve our goal or solve our problem?

II. WHO does NOT matter in the achievement of our goal or solving our problem?

III. WHAT does it do? WHAT does it NOT do?

IV. WHERE are all the places the problem surfaces or that our goal is difficult to achieve, and their characteristics? WHERE is the location for the ideal operating conditions (there is NO problem)?

V.   WHY is this an important goal to achieve or problem to solve? (What will happen if we do NOT address/solve this problem?)

VI. HOW do we know that the problem exists or the goal is difficult to achieve?

VII. HOW MUCH money/revenues do we believe we can save or make by solving our problem or achieving our goal?   15

Figure 8.4 shows the start of the TRIZ inventive principles Question Bank for the SMEs to use to address or overcome each constraint in #2 of Figure 8.3 to generate opportunities in #3 of Figure 8.3. Each of the opportunities or root WHYs generated in #3 of Figure 8.3 becomes an opportunity or conceptual direction for ideation.

Step 5: Generate a Question Bank

> Millions saw the apple fall, but Newton was the one who asked why.

> **Bernard Baruch**

Once you have filled out all the #2 and #3 steps in Figure 8.1, you are ready to use the Question Banking generation recipe to start creating open-ended, thought-provoking questions for ideation. The first step is to treat each combination of 1 + 2 + 3 in Figure 8.1 as triplets, and circle the triplets that are the most compelling or boldest conceptual directions to explore. This is done by reviewing the root #3s generated in Figure 8.1 and determining with the SMEs which opportunity space sounds the most promising to explore with a larger group. Sometimes the SMEs realize that they already have generated a solution at this stage and the Directed Innovation exercise can be considered completed. However, if there are compelling conceptual directions to explore that would be worth ideating with a diverse group of creative individuals, then the question-generation recipe can be used: *How might we use opportunity #3 to overcome limitation #2 and achieve/remove #1? Or, How might we achieve/remove #1 by using #3 without #2?* The facilitator walks through the most compelling triplets with the SMEs, turning each into an open-ended thought-provoking question.

The following are suggested wordsmiths and polishes of the questions:

1. Reference http://www.thesaurus.com to ensure that the simplest adjectives and nouns are used describing the *what* of the objective, without prescribing *how*. The question must be non-leading and not contain a potential answer or solution.
2. Increase or convert each question to be *open-ended* (eliminate *yes* or *no* questions).
3. Replace *can/could/should* with *might* and *may* so that ideators suspend disbelief to consider options that may stretch assumptions and laws of science.
4. Generalize so non-domain experts can relate and engage in ideation readily.
5. Tease out inflection points: conflicts, contradictions, and trade-offs—look for opportunities to frame questions as BOTH ... ANDs without any BUTs.

Then the facilitator and SMEs perform a quality review of all the questions, prioritizing them in the order of most compelling and most novel (a prior art search should be conducted on the content of the questions using Google Patents or freepatentsonline.com) to least, since the ideation session time allotment may not allow for addressing all the questions. The facilitator will help the primary SMEs draft questions that guide the ideators away from existing published solutions, patents, or prior art.

Quality Review Checklist for Ideation Questions

✓ Brief and concise
✓ Provocative, inviting, and inspiring
✓ Clear and focused
✓ Understandable by a variety of people
✓ Grammatically correct
✓ Functional, action-oriented verbs that describe the desired result or outcome

Once the questions are prioritized, they are grouped into categories or clusters addressing the same constraint or restriction, or exploring the same solution space. They are prioritized within each category and numbered. Then, each category of question is given a top level, broader question, and each subquestion is listed underneath it on one to three PowerPoint slides in preparation for the ideation session.

So you would have at most four big questions in 1 hour, with each of the one to four questions having sub-questions numbered 1.1 through 1.4, etc., and 2.1 through 2.4, etc. This is important for the documentation, sorting, and later homework given out to the ideators for tracking problems and solutions to closure. It is also important if patent filings eventually are made to track the original problem statement, the questions answered, and all the related or overlapping solutions that were generated so that a broader patent application can be filed.

Step 6: Select ~20 Diverse Participants

When selecting people to participate in the actual ideation session, you want to select at least half of the people to have some familiarity or expertise with the problem domain, and half with less familiarity and less expertise. Depending on the complexity of the problem and how much time you have to invest in simplifying the problem to engage nonexperts, the facilitator will need to determine the right mix and degree of expertise required to engage in ideation during planning. From experience, it is typically a mistake to only engage SMEs in the ideation phase. The best results come from having product managers, sales or marketing people team with engineering or operations staff.

It is best to create a team of ideators that is more than 8 and fewer than 30 both for diversity of thought and for one experienced facilitator to be able to manage and focus the group.

Step 7: Use Question Banks to Ideate in Pairs, 15 Minutes for Each Big Question

Ensure ideators can all focus on the same question at the same time in the same room; pairing ideators who have external customer understanding and knowledge with those who have an understanding of your company's operations and competencies works best.

Do not divide into small groups and send to breakout rooms working without a facilitator, or the focus of the ideators and both quantity and quality of solutions will suffer; there is a positive energy and enthusiasm created when all ideators are generating idea sheets and can look around the room and see how other pairs are faring. A competitive atmosphere can be created, and facilitators can leverage this to ensure all teams are producing idea sheets. An unleashing of creative potential occurs when people write down all their ideas, instead of self-filtering. The focus is on generating as many ideas as possible, and to get the ideators to suspend their critical thinking mind-set and find solace that all *bad* ideas will get filtered in a later convergent step in the Directed Innovation process. No one wants a *bad idea* tracked back to them as the originator. However, some of the *bad or incomplete ideas* generated at this juncture may combine with other ideas in the evaluation phase to produce an elegant and promising solution.

No one should teleconference into the ideation session; people working alone without a facilitator will start multitasking and lose focus on the question at hand or problem to be solved. Facilitator(s) have a role to ensure each team or pair is productive, focused, and capturing all of their ideas and writing them down, as well as drawing visuals depicting the process or architecture of their potential solution.

Never pair two SMEs. Have a group of three, with at least one person who is not internal or externally focused in the group, if you have an uneven number of participants.

Everyone will address the same problem (question on a projected PowerPoint slide) at the same time. The facilitator should rotate partners after every problem, and, at a minimum, once every hour. This way no one person is paired with someone who is having an off day. Also, the chemistry and diversity of experience will vary. Some people are naturally more creative and will engage their partners in ideation. The facilitator should walk the room ensuring that all pairs of ideators are working well together, are using their 15 minutes effectively, and are writing all their ideas down.

Otherwise, the ideators might discuss a lot of ideas but not have time to record any. It is critical (for follow-up and intellectual property protection purposes) that ideators act as their own scribes and record as many of their ideas in as much detail as possible on the idea sheets they are given (one idea per sheet), like the one in Figure 8.5. For legal purposes, each ideator should ensure their employee identification is clearly captured on each idea to which they substantively contributed.

If the ideators are very busy and continuing to generate robust solutions after 15 minutes, facilitators should allow the group to continue working on the same question until there is a lull in productive activity. It is best to also bring printouts of the Question Bank (e.g., PowerPoint deck of numbered questions), if there are some pairs who would like to move to the next question, and the majority of the ideation pairs are still generating robust solutions for the current problem/question. It is important to keep all pairs of ideators engaged, as the flow of ideas, once disrupted, is difficult to jumpstart again.

Very short 5-minute breaks should be given between question slides within the same problem domain. If the ideation session is 3–4 hours long, then there should be longer 10-minute breaks between each significant problem domain, and ideation partners should be switched during these breaks. It is a good idea to allow for breaks when people can physically move, since exercise and movement also helps with right-brain functioning.

Be sure to have plenty of right-brain-boosting snacks on hand for these breaks, such as chocolate, water, nuts, cinnamon, peppermint, and green tea.

Step 8: Combine, Evaluate, Eliminate, and Distribute Idea Sheets

Hopefully, you complete step 7 with more than 100 idea recorder sheets, all sorted by the facilitator into stacks by the question number each solution is meant to address. Tip: as the facilitator collects the idea sheets during step 7, he or she should also list out the total number of ideas generated so far on a white board or easel sheet, so that the participants can observe their great progress.

Step 8 is conducted with the core team that planned for the ideation event and generated the Provocation worksheet and Question Bank for use in ideation. It is recommended that a patent attorney or patent agent familiar with the problem domains also participate, if novel solutions are a goal of the exercise, and plans exist to commercialize the solution.

Session name: XXX Innovation Workshop		*Confidential when Completed*
What problem are you trying to solve? *(If working from a list of questions, record the question number.)*	**What is a "working title" or keywords for your innovation?**	
	How might your idea/solution be implemented? *(A sketch, flowchart, or list of features will help to explain this.)*	
What is your idea/solution?		
	Idea Recorder	
Innovator(s) D(s):	**Suggested lead:**	**Potential business value:** High, medium, low, unknown	**Today's date:**

FIGURE 8.5

Recorder sheet for capturing each idea.

1. Assign each big question or problem domain stack of idea sheets to a relevant SME. Have the SMEs first go through and combine any similar ideas/solutions with paperclips.

2. Once each SME has completed the combination task, have them eliminate any idea sheets that are not robust potential solutions. Owing to the nature of the ideation exercise, these should be obvious where the ideators were simply writing to keep going, but did not record a valid or complete concept. Some of these partial solutions may be combined with other robust solutions in the first step.

3. All remaining solutions are evaluated for whether they actually solve the problem or answer the question, and their relative impact on the customer or end user's operations if implemented. A team can develop and use their own success criteria for rating and selecting solutions.

4. A patent attorney or agent can then eliminate any of the solutions or ideas that the company or competition have already commercialized or patented—from their own knowledge of the domain. Some SMEs are very good at performing this elimination exercise on their own.

5. The remaining concepts and ideas are numbered and captured, with relevant inventor IDs, into a spreadsheet and scanned into a linked document repository (e.g., Autonomy, SharePoint, Livelink, etc.), and made accessible to the entire ideation team for further processing in step 9.

Step 9: Generate Metrics, Mentor Innovators, Track Ideas to Closure

Step 9 requires tenacity, discipline, attention to detail, and influence skills. The most frequent reason given for the failure of Directed Innovation is the lack of implementation of step 9. Ideation is fun! This is the hard work, nose-to-grindstone step of the overall method. This step comes when all the ideators have returned to their *day jobs* and the ideation activity is the last thing on their mind.

Checklist for implementing step 9 effectively and efficiently:

1. A project manager is assigned to manage the process in Directed Innovation step 1. The facilitator is agnostic, as you may recall, and should not be the in-business owner or project manager to ensure closure on all Directed Innovation activities.

2. Project management responsibilities can be distributed across each problem domain. If there is senior management sponsorship for each problem domain, from activities conducted in Directed Innovation step 1, then generating an intellectual property portfolio around each solution space should be on each project manager's radar, from the sponsoring organization's management.

3. The recommended approach for project manager(s) is Plan–Do–Check–Act for each idea that moves forward from step 8. Assign a lead inventor/ideator from those listed as co-inventors of the final combined concept in the list generated at the end of step 8. Assign an inventor mentor to each of these ideas. Inventor mentors are those skilled in conducting prior art searches for novelty, and enablement of patent disclosure write-ups. Ensure that the lead inventor and the inventor mentor work with the rest of the inventor team for each idea to transform it into a complete solution addressing the original problem or question. Once a detailed, enabled implementation is documented (e.g., disclosure form generated), the inventor mentor and lead inventor can conduct a prior art search to ensure that the solution is not already patented. This will ensure freedom of action if the company decides to take this solution to market.

4. Project manager(s) should regularly report to sponsors on the status of ideas implemented in products, as well as generation of patent filings or trade secrets.

5. Project managers should identify opportunities for further Directed Innovation by analysis of those questions for which no viable solutions of value to the company or the customer were generated. They should also be on the lookout for new gnarly problems that would benefit from a Directed Innovation session.

Addendum: Necessary Skills for Successful Directed Innovation Implementation

Critical facilitation and inventor mentoring skills can make or break the success of your Directed Innovation session.

Directed Innovation Facilitator Skills

- Effective group facilitation and meeting management
- Problem identification and analysis—analogous and adaptive thinker

- Broad technical knowledge of your company's verticals and adjacencies
- Process-oriented and disciplined time manager
- Well-read: trend watcher and technology scout
- Excellent communication and presentation skills
- Motivational and enthusiastic
- Tutoring/instructional background a plus

Inventor Mentoring Skills

- Creative, tenacious problem solver
- Collaborative coach
- Competent investigator
- Flexible, resilient, and adaptable
- Technical experience in the relevant field
- Address relevant art barriers to novelty
- Address enablement/reduction to practice issues
- Advocate/shepherd promising inventions

EXAMPLE

Conducting steps 3 through 5 of Directed Innovation (see Figures 8.6 through 8.10).

From observations to
insights/contradictions

1. You observe a stressful situation or problematic task.
2. You or the customer identifies a need or a potential fix or a process improvement.
3. Write it down: *"We want more closet space."*
4. Now, identify at least one of the obstacles to doing that—from your observations or interviews.
5. Write that down: *"If we only keep one purse, we give up style and color-coordination."*
6. Rewrite the contradiction with an inventor's mindset: *"We want BOTH matching purses AND plenty of storage space."*
7. Now don't dismiss it! Park on it! Ponder it! With **Provocation**
8. **DI** => Find a solution that "**resolves** the contradiction."

FIGURE 8.6
Process.

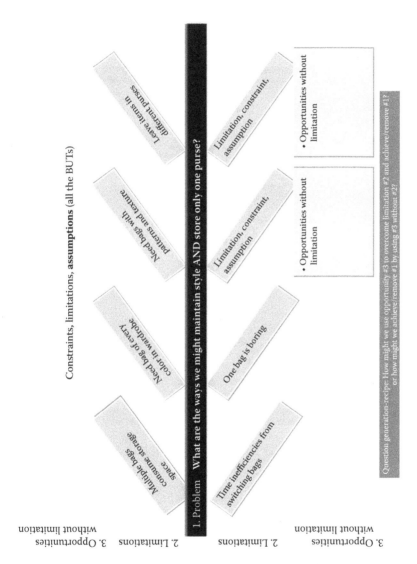

Constraints, limitations, **assumptions** (all the BUTs)

3. Opportunities without limitation

2. Limitations

1. Problem   **What are the ways we might maintain style AND store only one purse?**

2. Limitations

3. Opportunities without limitation

Leave items in different purses

Need bags with patterns and texture

Need bag of every color in wardrobe

Multiple bags consume storage space

Limitation, constraint, assumption

Limitation, constraint, assumption

One bag is boring

Time inefficiencies from switching bags

• Opportunities without limitation

• Opportunities without limitation

Question generation-recipe: How might we use opportunity #3 to overcome limitation #2 and achieve/remove #1? or how might we achieve/remove #1 by using #3 without #2?

FIGURE 8.7
Constraints and limitations.

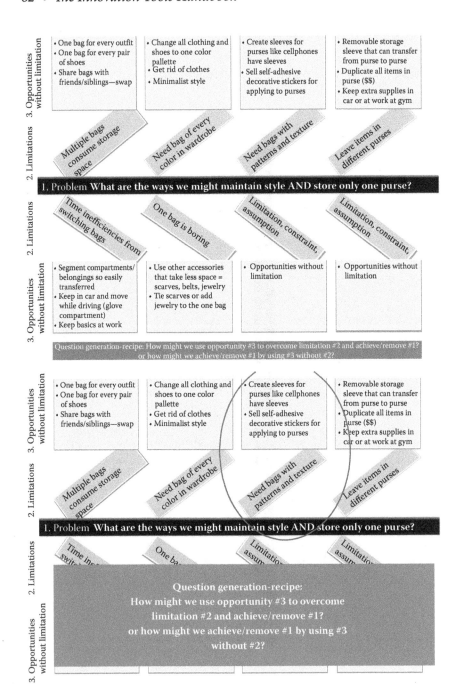

FIGURE 8.8
Recipe.

1. What are all the ways we might maintain style AND store only one purse?

    1.1 What are all the ways we might **have bags with multiple colors and textures** AND only one purse to store?

        1.1.1 What are all the ways we might create sleeves or covers for purses (like cellphones have sleeves)?

        1.1.2 What are all the possible attachment mechanisms for these sleeves or covers?

        1.1.3 What are all the types of covers we might create?

FIGURE 8.9
Outcomes and limitations.

1. What are all the ways we might maintain style AND store only one purse?

    1.2 What are all the ways we might **create the illusion of a different purse to match every outfit or shoes**?

        1.2.1 What are all the patterns that are on trend for clothes and shoes?

        1.2.2 What are all the styles for purses that align with clothing and shoe styles (e.g., casual, business, fun, play, beach, conservative, evening wear)?

        1.2.3 How might we make purse styles easy to collect, swap, share and trade with friends?

        1.2.4 How might we have all these styles and consumes minimal storage space in our closet?

FIGURE 8.10
Results.

SOFTWARE

- Innovation Problem Solving—Ideation: http://www.Ideationtriz.com

REFERENCES

Bailey, C., 2013, 9 Brain foods that will improve your focus and concentration. Available at http://ayearofproductivity.com/9-brain-foods-that-will-boost-your-ability-to-focus/.

Drew, J., 2011, 50 Ways to Boost Your Brain Power. Available at http://wakeup-world.com /2011/06/02/50-ways-to-boost-your-brain-power/.

de Bono, E., Training for business: Lateral Thinking, Directed Attention Thinking Tools, Six Thinking Hats. Available at http://edwdebono.com/debono/qtraina.htm.

Domb, E., *Global Innovation Science Handbook*, Chapter 24, TRIZ: Theory of Solving Inventive Problems, McGraw Hill Education, copyright 2014. Available at https://www.mhprofessional.com/product.php?isbn=0071802088.

Haman, G., Question Banking, available at: Gerald 'Solutionman' Haman talks about innovation challenges, creativity tools and creative workspaces @ http://www.innovation management.se/imtool-articles/gerald-solutionman-haman-talks-about-innovation -challenges-creativity-tools-and-creative-workspaces/.

Lauchner, C., Assumption Storming, Front End of Innovation Conference, April 7, 2014. Available at https://prezi.com/k3g2i_furntp/copy-of-fei-2014-assumption-storming/.

Miche Bags, Build Your Bag internet shopping site available at http://miche.com/.

Williams, I., Frog Design, November 10, 2011, The Art of Provocation. Available at http://www .core77.com/posts/20995/drc-x-2011-design-research-for-creating-change-20995.

SUGGESTED ADDITIONAL READING

Bens, I., 2012. *Pocket Guide to Facilitation*. GoalQPC.

Domb, E. and Rantanen, K., 2008. *Simplified TRIZ: New Problem Solving Applications for Engineers and Manufacturing Professionals*, 2nd Ed., Taylor & Francis Group. ISBN-13: 978-1420062731.

Gelb, M.J. and Caldicott, S.M., 2008. *Innovate Like Edison*. Penguin Group. ISBN-13: 978-0-525-95031-8.

Lauchner, C. Father of the Assumption Storming ideation method, website at http://assumptionstorming.com/.

Miller, C. Innovating Results! Website containing 40 Inventive Principles TRIZ question bank. Available at http://innovatingresults.com/uploads/TRIZ_QuestionBank.pdf.

Strachan, D., 2007. *Making Questions Work: A Guide to How and What to Ask for Facilitators, Consultants, Managers, Coaches, and Educators*. 1st Ed., Jossey-Bass. ISBN-13: 978-0787987275.

9

Elevator Speech/The Coffee Clutch Opportunity

H. James Harrington

CONTENTS

It is better to have a productive 30 seconds with the right person than two hours with most people.

DEFINITION

- Elevator speech—An elevator speech is a clear, brief message or *commercial* about an innovative idea you are in the process of implementing. It communicates what it is, what you are looking for, and how it can benefit a company or organization. It is typically no more than two minutes, or the time it takes people to ride from the top to the bottom of a building in an elevator.

- Coffee clutch opportunity—The coffee clutch presentation is used primarily in a relaxed environment and consists of a very short discussion that lasts between 90 and 120 seconds. It typically takes place around the water fountain, on the walk to or from a meeting, or during a coffee break.

USER

This tool can be used by individuals who are trying to convince an individual or group to become supportive or actively involved in their part of the innovative process. A variety of people, including project managers, salespeople, evangelists, and policy makers, commonly rehearse and use elevator pitches to get their point across quickly.

OFTEN USED IN THE FOLLOWING PHASES OF THE INNOVATIVE PROCESS

The following are the seven phases of the innovative cycle. An X after the phase name indicates that the tool/methodology is used during that specific phase.

- Creation phase
- Value proposition phase X
- Resourcing phase X
- Documentation phase
- Production phase X
- Sales/delivery phase X
- Performance analysis phase

TOOL ACTIVITY BY PHASE

- Value proposition phase—The elevator speech is used during the value proposition phase to provide management with an overall understanding about the innovative concept and how it will affect the organization.

- Resourcing phase—The elevator speech is used to acquaint management with an understanding of why they should invest resources into the proposed innovative concept.
- Production phase—The elevator speech is used to help get the affected people to participate in the changes necessary to implement the innovative concepts.
- Sales/delivery phase—The elevator speech is used to capture the attention of potential customers/consumers. It is also used to gain the enthusiasm of the sales force related to the innovative concept.

HOW TO USE THE TOOL

An elevator speech is a clear, brief message or *commercial* about the innovative idea you are in the process of implementing. It communicates what it is, what you are looking for, and how it can benefit a company or organization. It is typically no more than two minutes, or the time it takes people to ride from the top to the bottom of a building in an elevator. It should be approximately 80 to 90 words or no more than 8 to 10 sentences. (The idea behind having an elevator speech is that you are prepared to share this information with anyone, at any time, even in an elevator.) It is important to have your speech memorized and practiced. Often two to four different elevator speeches are prepared for the same innovative concept, each focusing on the unique interest of the different individuals (e.g., investor, customer, management, or affected individuals within the organization).

The coffee clutch presentation is used primarily in a relaxed environment and consists of a very short discussion that lasts between 90 and 120 seconds. They are very similar to the elevator speech in intent and content, but the presenter has more time to convey the information to the audience. Frequently, these short overviews are also used as the introduction to a formal presentation.

Elevator speeches are supposed to grab investors' attention in less than a minute. What can you possibly say in such a short time that will make investors (and customers) want to hear more? The remainder of this chapter will try to give you some ideas and examples that will help you make your own elevator speech.

The name *elevator speech* reflects the idea that it should be possible to deliver the summary—or short speech—in the time span of an elevator ride, or approximately 30 seconds to 2 minutes, and is widely credited to Ilene Rosenzweig and Michael Caruso—while he was editor for *Vanity Fair*—for its origin (Pincus, 2007).

The term itself comes from a scenario of an accidental meeting with someone important in the elevator. If the conversation inside the elevator in those few seconds is interesting and value adding, the conversation will continue after the elevator ride or end in the exchange of business cards or a scheduled meeting.

The elevator speech and the coffee clutch opportunity are carefully prepared presentations that focus on providing the maximum amount of relative information in a minimum amount of time, and is as simple as possible. Often, they are prepared for specific audiences. For example, if you are talking to a potential investor, you may want the focus to be on return on investment. If you are talking to the people that will be affected by the implementation of the innovative concept, you may want to emphasize how it will make their job easier and more interesting. If you are talking to the consumer, you would focus on how the new innovative concept would be of value to them.

Four Parts of the Elevator Speech

In preparing the presentation, we find it is useful to think of it in four parts:

1. What is the unfulfilled need and why is it important to fill that need?
2. What is the innovative concept?
3. What are the benefits?
4. What does it take to implement an innovative concept?

Sample Elevator Speech Outline

The following 10 speech topics will help to write a carefully planned and prepared presentation that grabs attention and says a lot in a few words. This format suggestion helps you avoid creating a sales pitch. Use each idea to write one short powerful sentence.

1. Smile at your counterpart, and open with a statement or question that grabs attention: a hook that prompts your listener to ask questions.

2. **ABOUT YOU.** Tell who you are; describe you and your company.
3. Tell what you do and show enthusiasm.
4. **WHAT DO YOU OFFER?** Tell what problems you have solved or contributions you have made.
5. Offer a vivid example.
6. Tell why you are interested in your listener.
7. **WHAT ARE THE BENEFITS?** Tell what very special service, product, or solutions you can offer him or her.
8. **WHAT ARE THE BENEFITS?** What are the advantages of working with you? In what way do you differ from competitive companies?
9. **HOW DO YOU DO IT?** Give a concrete example or tell a short story, show your uniqueness, and provide illustrations on how you work.
10. **CALL FOR ACTION.** What is the most wanted response after your elevator speech? Do you want a business card, a referral, or an appointment for a presentation after your elevator speech?

Six Questions to Be Answered

When creating the elevator speech, there are six questions it must answer:

1. *What is your product or service?* Briefly describe what it is you sell. Do not go into excruciating detail.
2. *Who is your market?* Briefly discuss who you are selling the product or service to. What industry is it? How large of a market do they represent?
3. *What is your revenue model?* More simply, how do you expect to make money?
4. *Who is behind the company?* "Bet on the jockey, not the horse" is a familiar saying among investors. Tell them a little about you and your team's background and achievements. If you have a strong advisory board, tell them who they are and what they have accomplished.
5. *Who is your competition?* Do not have any? Think again. Briefly discuss who they are and what they have accomplished. Successful competition is an advantage—they are proof your business model or concept work.
6. *What is your competitive advantage?* Simply being in an industry with successful competitors is not enough. You need to effectively communicate how your company is different and why you have an advantage over the competition. For example, a better distribution channel? Key partners? Proprietary technology?

Tips on How to Do It

1. Give a concrete example or tell a short story, show your uniqueness, and provide illustrations on how you work.
2. Include a loud CALL FOR ACTION.
3. What is the most wanted response after your elevator speech? Do you want a business card, a referral, or an appointment for a presentation after your elevator speech?
4. You may want to keep small take-away items with you, which you can give to people after you have delivered your pitch. For example, these could be business cards or brochures that talk about your product idea or business.
5. Remember to tailor your pitch for different audiences, if appropriate.
6. These are other points, questions, and business subjects you could ask yourself:
   - Who is your target?
   - How large is your market volume?
   - How do you make profits?
   - What are the background, major milestones, and achievements of your team?
   - Who are your competitors? How do they solve a problem?
   - What is your strength and advantage compared with them?
   - What is your unique selling proposition?
   - What are their special patents or technology?
   - Do you have a special approach in client management?
   - And so on.

Checklist for Fine Tuning (Pincus, 2007)

Step 1: First write down all that comes up in your mind.

Step 2: Then cut the jargon and details. Make strong short and powerful sentences. Eliminate unnecessary words.

Step 3: Connect the phrases to each other. Your elevator address has to flow naturally and smoothly. Do not rush.

Step 4: Memorize key points and practice.

Step 5: Have you really answered the key question of your listener: What is in it for me?

Step 6: Create different versions of your elevator speech for different business situations. Note them on professional business cards.

EXAMPLES OF AN ELEVATOR SPEECH

1. Imagine that you are creating an elevator pitch that describes what your company does. You plan to use it at networking events. You could say, "My company writes mobile device applications for other businesses." But that is not very memorable! A better explanation would be, "My company develops mobile applications that businesses use to train their staff remotely. This results in a big increase in efficiency for an organization's managers." That is much more interesting, and shows the value that you provide to these organizations.

2. Imagine you are creating an elevator speech to convince the organization that they should implement a special program focused on improving innovation throughout the organization.

First Version: Install Innovation System Elevator Speech

Have you seen the latest announcement that Google made about their new computer chip? It is 100 times faster than anything that has been produced to date.

You know they are attracting the best and the brightest. Three out of the five interns from Stanford that we made offers to went with Google and another went with IBM. Personnel contacted all four of them, and it was not the money, but the challenge, that made the difference. We need to improve our level of innovation to become a leader, not a follower. I found studies that show that organizations that have active projects to increase innovation levels are 20% to 30% more productive.

There is an excellent study that will measure our innovation level in the key parts of our organization. This will help us compare our innovation level to other organizations and point out improvement opportunities. It is particularly directed at research and development, our management systems, production, sales, and marketing. Personnel has informed me that our creativity activities have been primarily focused on research and development. I agree with them that research and development innovation is important, but I also believe that we have to have all parts of our organization actively searching for new ways to be innovative related to the way we do business, the way the individual functions do business, and how well we work together. Do you agree?

Can I call you and set up a meeting where we can talk about this in more detail? I have run across some case studies that are really very impressive that I would like to share with you.

Second Version: Install Innovation System Elevator Speech

Have you seen the latest announcement that Google has made about their new computer chip? It is 100 times faster than anything that has been produced to date. You know they are attracting the best and the brightest. Three out of the five interns from Stanford that we made offers to went with Google and another went with IBM. Personnel contacted all four of them, and it was not the money, but the challenge, that made the difference. I have found studies that show that organizations that have active projects to increase innovation levels are 20% to 30% more productive.

There is an excellent study that will measure our innovation level in the key parts of our organization. This will help us compare our innovation level to other organizations and point out improvement opportunities.

Can I call you and set up a meeting where we can talk about this in more detail?

SOFTWARE

No software is required for this tool.

REFERENCE

Pincus, A. The perfect (elevator) pitch. *BusinessWeek*, June 18, 2007.

SUGGESTED ADDITIONAL READING

Hahn, G.J. *Statistics-Aided Manufacturing: A Look Into the Future, The American Statistician* vol. 43, no. 2, pp. 74–79, 1989.
Peters, T. *The Wow Project, Fast Company*, vol. 24, p. 116, 1999.

10

Ethnography

Brett Trusko

CONTENTS

DEFINITION

Ethnography can be used in many ways, but most significantly in the creation of a new product or service with a clear understanding of the many different ways that a person may accomplish a task based on their own worldview. Ethnography is observing and recording what people do to solve a problem, and not what they say the problems are. It is based on anthropology but used on current human activities. It is based on the belief that what people do can be more reliable than what they say.

USER

This tool can be used by individuals, but its best use is with group that can empathize with a larger group (racial, cultural, lifestyle, etc.) that may be different from the traditional customers or the individuals who work for the company. As an example, an individual based in one region of a country may not sufficiently understand the priorities in another part of the same country.

OFTEN USED IN THE FOLLOWING PHASES OF THE INNOVATIVE PROCESS

The following are the seven phases of the innovative cycle. An X after the phase name indicates that the tool/methodology is used during that specific phase.

- Creation phase X
- Value proposition phase X
- Resourcing phase
- Documentation phase
- Production phase
- Sales/delivery phase X
- Performance analysis phase

TOOL ACTIVITY BY PHASE

- In the creation phase, ethnographic tools are used to better understand the customers for whom the product will be built.
- In the value proposition phase, the tool may be used as justification for creating a new or different product or service.
- In the sales/delivery phase, ethnographic tools would be used to craft custom messages for certain groups of people, countries, and cultures.

HOW TO USE THE TOOL

Ethnography is an element of anthropology dealing with the systematic study of cultures and human behavior. Conventionally, anthropologists would spend time studying or observing patterns of human behavior. Today, ethnography has become a consumer research method for uncovering unspoken customer needs or to gain insights into customer desires for specific positive experiences. In the field of innovation, ethnography has taken on a meaning that is specific to business and can be applied to a single company in the understanding of the differences of people inside the organization, or to multiple companies or industries where one would look at the differences between organizations to understand them or to better work with another organization in a collaborative effort. In other words, ethnography can be used as an approach to break down silos. There are many people who are interested in innovation who argue that a requirement of innovation is tolerance. Tolerance, in the traditional sense the word, would mean that an individual or group of individuals would tolerate people with different ideas. In the raw sense of the word, this might mean tolerance of other ethnic groups or sexual preferences. The theory is that tolerance of others directly relates to tolerance of new ideas, and tolerance of new ideas leads to innovation. In Wild's (2012) article, "The making of water cooler logic stakeholder ethnography—Composting is a metaphor for innovation," she describes in a rather unorthodox way how innovation is the composting of many different ideas, attitudes, and approaches into a single new and innovative idea (the compost). From the perspective of ethnography, this is an excellent visual. Imagine, if you will, that you take the great melting pot that is called the United States, throw in all the different races, cultures, foods, etc., and stir them into a compost. The result of the completed process is the growth of new life (plants/ideas).

An interesting description really comes from the work of philosophers Barwise and Perry (1983), which describes a key obstacle in the relation of linguistic expression to worldly reference. For example, deictic references that did not name their objects but pointed to them from within a context, created problems. Expressions such as these could be *this* or *there*, and could only succeed in reference if the context was known and shared by everyone. The example he used was of a person holding a sign, a sign that reads, "I am a poor deaf–mute person. Can you spare a quarter?" changes its meaning depending on who is holding the sign. Therefore, the true value of the statement can only be assessed if the context is taken into account. In acknowledging these problems, you have a problem with situational semantics and logical grammar, attempting to embed the influence of the situation into its rules and truth tables. This relates to ethnography and innovation because people see the world differently depending on the context. In other words, people may use the same tool in different ways. To understand the context of a person or group of people, one must find a way to either understand it logically or embed themselves inside the culture of the user. In the case of the poor deaf–mute person, the non-deaf–mute might assume the poor refers to the feeling (feeling sorry for myself), while the true context may mean they do not have much money.

Another way in which ethnography is used in the business place is in the study of how people use something. In the early 1990s, Xerox Corporation did research on the interaction between copiers and human operators with the objective of making the design of the machine accommodate the manner of human reasoning. They carried out observations of real-time interactions between users and machines, and analyzed what people actually did and said instead of relying on the predictions of a model, which made it a major *ah ha* in the research world. The results of this research influenced the design of the copier–user interface to help the user to clear simple instances of malfunctioning equipment. This research was also used to develop a knowledge management system for repair personnel: using simple communications technology, repairmen were able to cooperate and share useful approaches for diagnosing and treating copier problems. Everyone at Xerox who expected to be dazzled by the latest and greatest technology used a very primitive communication device and relied on the collaborative practice of a professional community to diagnose problems, devise solutions, and accrue the knowledge to deal with the emerging high-end copier. Of course, the cost of repairing copiers benefits Xerox, and the end users of Xerox products. Use of ethnographic research became a banner item for Xerox. So, when you complement the interface on the computer, you are actually recognizing ethnographic research.

Given the growing demand for innovation, the role of exciting user experiences in the success of new products, and the role of design innovations in making user-friendly products, ethnographic research has become the go-to tool for driving product innovation.

In his book *Ethnography*, John Brewer states

> Ethnography is the study of people in naturally occurring settings or "fields" by methods of data collection that capture their social meanings and ordinary activities, involving the researcher participating directly in the setting, if not also the activities, in order to collect data in a systematic manner but without meaning being imposed on them externally. [sic]

Note that this definition works well because ethnography is about people. What is sometimes misunderstood about ethnography is that it is not always about other cultures, but could be about different cultures inside your organization. Not only can this apply to inside your organization, but it could also apply to your community, whether that community is of businesses or people. In other words, ethnography could be applied to the study of highly innovative organizations. This assumes that a highly innovative organization has a culture that may be different from yours. To emulate that culture, you must learn to understand how and why it functions as it does.

As mentioned earlier in this chapter, ethnography need not be about studying cultures in far-off lands; it can also relate to the layout of your supply cabinet in your office, or with the way a grocery store in one community might differ from a grocery store in another community. A simple example might be in the layout of sporting goods stores in San Francisco versus Houston, Texas. In San Francisco, you might expect to see running clothes and other fitness items more prominently displayed than hunting clothes. A sporting-goods store in Houston, Texas, is likely to have a hunting department take up a far greater percent of square feet. Of course, this is a stereotype, but since I have lived in both cities, I can say that it is true. The culture in Houston, Texas, is much more hunting and gun oriented than that in San Francisco. Once again, depending on your worldview, you may be shocked by (a) so much gun space or (b) so little. Obviously, this probably has less to do with the space provided than it does with the attitude of the people. If you were to apply this to certain types of restaurants or health clubs, you might find similar differences. In the area of innovation, for instance, I have heard over and over again from communities struggling to develop an innovation hub that the only problem is the lack of venture capitalists coming to the area. Of course, anyone who has spent a significant amount of time around innovators understands that innovation happens first and financing second. Once again, people outside the innovation community may not understand the perspective from inside the innovation community.

Ethnography is more of a qualitative research methodology that has its own challenges surrounding accuracy, validity, and reliability due to the natural settings of the research, participants' activities and choices, design of the research, and selection of participants. Ethnographic research explores the beliefs, practices, artifacts, folk knowledge, and behaviors of a group of people (LeCompte and Goetz, 1982). Unfortunately, the line between anthropology in social studies sometimes is not well understood by business people. Therefore, there are probably not as many ethnography tools in the research and practical application of ethnography to business situations, and product development remains an area ripe for additional work.

Need for Ethnography

The primary purpose of ethnography is to document or discover a faithful and accurate understanding of participants' ways of using certain methods or tools. Ethnography reveals subjective and subtle information about the behaviors of a target group of people that are difficult to identify in other scientific approaches to collecting data. As in any social science experiments, information gathered through ethnographic research is subject to known uncertainties surrounding individual behaviors. Also called field research, it is used to understand people's actions and their related and routine experiences. The methodology of such research is best summarized below:

> Field research involves the study of real-life situations. Field researchers therefore observe people in the settings in which they live, and participate in their day-to-day activities. The methods that can be used in these studies are unstructured, flexible, and open-ended. (Burgess, 1984)

As mentioned earlier in the chapter, the spectrum of research can extend from ethnography within your own home to a study of different countries. This expansion of what many consider to be classical ethnography demonstrates that ethnographic researchers must become intimately familiar with the people being studied as well as their environments, activities, and situations. Ethnographers sometimes become a subject of the study itself to gain deeper insights, its relevance, and meaning. Ordinary activities are studied without making it a planned experiment with unique data collection methods, and unstructured observations form the basis for analysis. At the end of the research, ethnographers may tell the story of their experience, present an analysis of the information gathered, and make their own. This can be particularly challenging for most human beings. Just consider some of the conversations you may have had with your relatives at the holiday table. The movie version tends to point out the absurdity and variance of opinions about so much of life all within a single family. If you have ever had an argument with your crazy relatives about politics or religion and cannot understand why they would think the way they do, you just discovered the challenge of ethnography. Being able to take yourself out of the situation and objectively observe and consider the routines, thoughts, perspectives, and worldview of people other than yourself is difficult for even the most open-minded.

In the business environment, ethnography is used as a deliberate inquiry process (Erickson, 1984) for the specific purpose of assessing the use of a certain product, improving the use of a product, or for creating a new product. It is not meant to be a formal, academic, and scientific research method to prove a hypothesis. Instead, the process is generally purpose driven, as in the example of the interface with the copy machine. This should not be confused with usability because usability is generally considered a constant. For example, if you were to design the cockpit of an airplane, you would probably design this reusability—the ability to reach certain switches and buttons in a certain sequence or under certain circumstances. This becomes an ethnographic problem when certain colors mean different things between cultures. As an example, the color green in the Western world is generally recognized, but green in Indonesia is seen as a forbidden color. Additionally, certain symbols may be a problem, which is why throughout the world, standardized international symbols have been developed for things like the restroom.

Some famous examples from marketing include the following:

- In 2007, the Cartoon Network launched a campaign that utilized a robot lit by light-emitting diode (LED) lights and placed them randomly throughout the United States. In Boston, someone thought the blinking LED device was a bomb and the city went into high alert, shutting down bridges subways and roads. The designers of the campaign were obviously insensitive to public sentiment as it related to blinking boxes. To anyone who spent a significant amount of time in a large East Coast city, especially after 9/11, one always remains vigilant of something that does not seem right.

- In 2011, Kenneth Cole, an international men's retail store, tweeted during the uprising in Egypt. The tweet was as follows: "Millions are in an uproar in #Cairo. Rumor is that they heard our new spring collection is now available online at xxx." In hindsight, this was both tacky and offensive to certain people.

- In October 2010, Gap clothing company ditched its new logo after only one week, due to an online backlash. Gap, which is known for its basics, was trying to present a more hip and upscale image. What Gap failed to realize was that its customers were more interested in basics versus trendy clothes. Sure, understanding exactly what their customers would feel about the new logo is always tricky; however, this is the marketing department's job—to understand their customers.
- In 2011, when Netflix attempted to launch Quickster as a way to promote streaming, their first attempt failed. Apparently, they did not understand the price point at which their customers felt DVD rental and streaming crossed. Also, a pot-smoking philosopher already owned the twitter handle.
- A classic case study for MBA programs everywhere is that of the 1985 New Coke fiasco. When Coca-Cola introduced a new formula to combat sweeter products, they had no idea how attached customers were to the traditional, all-American Coca-Cola brand.
- Not to pick on just Coke, Pepsi's launch into China was another fiasco. Pepsi's slogan of "Pepsi brings you back to life" translated in Chinese to "Pepsi brings your ancestors back to life." Had Pepsi done an ethnographic study, they would have understood the culture and not made such a serious mistake.

Researchers may have a set of questions that they want answered about what, how, why (or not), when, where, and how well people will use new products. Observed user reactions and experiences are captured and examined to establish meaning or the internalization of the user experience by the research subjects.

Meaning is commonly manifested in several ways: the removal of pain, connection to cultural practices, demographic preferences, relationships to other activities, changes in behavior, positive or negative impacts of interaction, and benefits of the experience. The more routine the behavior or benefit is to the user, the stronger the meaning of the experience will be. Ultimately, this information is used to create products that will provide enjoyable experiences and prevent painful challenges.

Ethnography and Innovation

Innovation can happen in form or function, or both. Form innovations are called design innovations and are based on more qualitative research, including ethnographic research. Typically, research leads to new opportunities for form or design innovations, which at times lead to the development of new technologies. For example, when the mobile phone was initially designed by technologists with an old calculator-type keypad, the marketing team rejected the idea because of its poor human interface and recommended instead the familiar phone-type keypad for better acceptance by users. Of course, for most business people who are used to using a keypad, it would have made sense and then easier to dial utilizing that layout. Obviously, the decision to go with the keypad as it is currently used was the right one.

Ethnographic research tends to be most effective for *platform*-type innovations—more so than for fundamental, derivative, or variation types where individual users would play a bigger role. For platform innovation, ethnographic research is a compelling form of co-creation.

We know that technology innovations increase a company's revenue, and we have learned that when design innovation is added to the mix, that revenue benefit is kicked up a notch or two. Enter ethnographic research! Other uses in ethnography in innovation are utilizing the methods to find out what target groups might actually want. As discussed earlier in the chapter, managers, research and development professionals, marketers, etc., may not always understand the needs of the customer. A somewhat notable example of this process in action is in the 2000 film *What Women Want*, where a chauvinistic ad executive accidently *goes native* and discovers what it is that women really want in order to win an ad campaign from a female executive. Although his original motivation is purely about beating a woman out of the ad campaign, he begins to understand them better and soon becomes enlightened. Imagine what a company might garner if they were able to really understand markets in which they compete.

Ethnography Skills and Tools

Ethnography is a holistic approach to studying a cultural system where the culture can change from place to place, or one market segment to another. Culture, a soft attribute, may include physical environment, historical behaviors, or traditions shared within a societal segment, and perceived priorities. All of these aspects evolved over time. Thus, ethnographic research is scoped by market segment and time in space. We begin the process by establishing a framework for study and outlining the purpose of the research and the intended target objectives. Remember also that with new tools such as big data analytics, we can begin to understand culture more concretely. For example, Spelunk utilized its *Spelunk for Good* platform to analyze tweets during Hurricane Sandy in New York. By doing this, they immediately knew where electricity was out, and where there were problems with flooding. Imagine if they went back to look at this data again to see how certain neighborhoods, and ethnic and socio-economic groups responded to the hurricane. It seems that this could be done for almost any question where different groups want to be understood. Of course, the limitation by only analyzing tweets or people who are computer users would not get to an academic journal, but would probably work really well for the marketing department.

Similar to other professions such as auditors or research scientists, ethnographers must have superior curiosity and the determination to dig deeper for information. They must be good at collecting available information, understanding what is not available, and thinking quickly of alternative sources. Strong communication skills are required to conduct thorough and probing interviews in a friendly and nonthreatening way, and they must be able to analyze information to recognize patterns. Moreover, an ethnographer must plan the research; understand the cultural aspects of the group, organization, or society under study; and summarize findings in an easily understood form for use by the target audience. This generally includes product or service designers. Additionally, as alluded to earlier in this chapter, ethnographers must be very careful to consider all options. Just as the arguments over the family dinner table can be caused by differences in opinion or differences in worldview, the ethnographer runs the risk of inserting his or her own bias into the research. Therefore, it can be challenging to do ethnography research unless the field team is highly trained.

During the observation phase, the focus is on recognizing and recording obvious patterns, giving attention to minute details of those activities that convey the meaning or have significant impact on the overall experi-ence of the subjects—both implicit and explicit. Implicit expectations may be manifested through facial and verbal expressions (such as frown, smile, curse, complaint, compliment, pain, and pleasure); demonstrated need for extra effort (such as excessive time to perform, extra effort to push, extra movement, proneness to mistakes or getting hurt); unexpected or unintended ways of doing things; and requiring additional resources for certain activities. The deeper the understanding of human behavior and its relationship to the user experience, the more opportunities to discover new ways to achieve breakthrough innovations.

Using the right analytical tools is equally important to making insight-ful observations. Graphical tools for capturing patterns, descriptive sta-tistical tools to summarize observations, and inferential statistical tools for comparative analysis can be used to address sample size limitations and reduce the risk of drawing incorrect conclusions. Many ethnography-specific software applications are available for gathering and analyzing data. Throughout the research, ethnographers must remain vigilant to recognize what else needs be collected. Tools such as those mentioned ear-lier offer tremendous opportunity in this phase of the project.

Interviewing is the next critical step for capturing and validating relevant information. A healthy respect and admiration for the culture of the group that is being studied underpins successful research by helping the ethnog-rapher sort between the positives and negatives inherent in any group.

During the interviews, ethnographers must remain mindful of local customs, sensitivities, sensibilities, and conversation protocols. Leading with open-ended questions will prevent defensiveness and increase the quantity of information. Friendly and frank conversations lead to unbi-ased responses and more usable data. As appropriate, researchers may use audio and video recording tools so they can be analyzed thoroughly and repeatedly, if necessary.

All the data in the world would do no good if it is interpreted out of con-text. Interpretation must be bound by the purpose of the research, the cul-ture of the group, and available data, but carefully interpreted and analyzed, they may lead to enjoyable user experiences. Ultimately, the innovation solution must work differently, look different, and offer different enjoyable experiences to create demand for the new product or service. The key being that a color, shape, or word that might be unappealing to an individual person, group, or culture may have the opposite effect on another group. An example of this in modern days can be seen by the attitudes toward sex as you move from one country to the next. In some countries, television will show partial or full nudity and in other countries nudity is completely taboo. While your product may have nothing to do with sex, there are still norms that may inspire you to innovate specifically for that market.

Speeding Ethnographic Research

Ethnographic research can be expedited without detriment to its power of insights. Similar to rapid prototyping and other rapid product development methods, rapid ethnography has also been proposed (Millen, 2000). A typical ethnographic research project that could take months to complete can be accelerated or shortened by applying other techniques to the tasks of identifying the subjects of research and observation, capturing information, and generating insights. Alternatively, utilizing big data analytics could, in theory, eliminate many of the steps requiring interviewing, etc. Caution should be taken, however, since analytics performed by people outside of the group being analyzed may result in something similar to garbage in, garbage out.

One such technique is the lead user concept (von Hippel, 1988) that has been used for identifying market opportunities by receiving feedback from early adopters of new solutions in the marketplace. Lead users are individuals who like to experiment with new technology or solutions, or they may have direct use for the new solutions. It can also be used for ethnographic research to identify subtle requirements based on their qualified experiences. Lead users could also be considered a *group* for purposes of ethnographic research. Once again, as stated earlier, the term ethnography may not be best since an ethnographic study need not have anything to do with ethnicity.

Instead of identifying a random sample of the group under study, the ethnographer would identify representative lead users within the selected group for getting information. It is like working with sample averages versus the population. In doing so, we may lose resolution or miss the finer inputs we could get from many individuals; however, this can be mitigated by more rigorous preparation for the ethnographic research project.

Narrowing the scope of the project will also shorten the overall time of the project. Instead of studying a social group in a target market for a larger purpose, we could quickly collect the information about specific aspects of the intended innovative solution. Instead of one large ethnographic research, we may take multiple short trips to the group for specific information. Thus, dividing and focusing on a smaller scope would accelerate the collection of input for new products.

Also, for greater efficiency, interviews and activities could be recorded in audio/video, and physical objects could be captured in high-resolution photos during the data collection phase. The resulting findings from analysis derived in isolation can be validated separately with the so-called smart people in the target group.

Organizing ethnographic research such that various aspects could take place in parallel could also reduce time to completion. As parallel processing has been used in manufacturing to reduce cycle time for producing products, performing ethnographic research with a select sample of informants to perform interviewing, data collection, and recording could occur simultaneously with more researchers for field research. Today, webcams or similar technology could be used to gather data remotely over a longer period of time, or even continually.

Institutionalizing Ethnographic Research

Considering the success of a few products that utilized ethnographic research for capturing unmet requirements, ethnography is becoming an integral and routine part of capturing and validating user requirements for new product development. To make it routine, it must become a standard process and be deployed economically and rapidly. Once again, even when using analytics on large samples of people, be mindful of the human part of the interaction.

Organizations may take a stab at establishing a standard operating procedure for ethnographic research (Venkatesh, 2001) using these steps:

1. Study people's behavior in natural settings rather than experimental ones.
2. Collect data specifically to identify unmet or hidden user expectations.
3. Be open-minded in collecting data in any form it is available rather than in a prescribed format.
4. Scope out the research based on the type of innovation, and target platform for the greatest benefit.
5. Take detailed notes from many perspectives while in the field. Capture exact terms used or views expressed by participants.
6. Select lead users to gather more information faster to shorten project duration.
7. Try to live the participant's experience in order to understand the emotional meaning of the experience.

8. While interviewing, let the participants express their responses and experiences in their own ways, terms, and expressions. Capture them all as is without questioning or formatting.
9. Identify multiple sources of information for a variety of data, and cross-validate findings.
10. Anticipate and prepare probing questions ahead of time, but be prepared for impromptu discussion.
11. Be a courteous, caring listener and express appreciation for participants' support.
12. Do not criticize any social group, its members, leaders, decisions, or activities.
13. Most important, have fun to bring out your own best.
14. Analyze, distill, and extract as many findings as possible, but pay more attention to unmet needs, subtle experiences, or insights.
15. Prepare a positive and supporting report.

Using the Tool

- Observe—Monitor how people perform certain activities in their natural environment.
- Interview—Talk to people to gain insight into how they accomplish their tasks.
- Survey—Through electronic and manual tools, gather answers to questions through surveys.
- Data analysis—Analyze data through manual and electronic tools.

SUMMARY

Ethnography is a research method for the social sciences. It is a relatively new process incorporated into the product development process for user-friendly innovation, to gain better acceptance and loyalty, and to capture greater market share. It involves the *up close and personal* participation of the researcher in a social group environment to capture hidden, unmet needs, and insights to make innovations more useful and to deliver a more enjoyable user experience. However, ethnographic research must make economic sense and be conducted proportionate to the type of innovation and market targeted. Literally, ethnography means a portrait of people

(Harris and Johnson, 2000). Interestingly, today's technology can help capture the finer portrait faster and even virtually.

EXAMPLES

There are few formal tools to work with in ethnography. Since ethnography is a qualitative science, the first three steps of observe, interview, and survey differ from situation to situation.

There are many examples where ethnographic research was conducted during the product development phase that resulted in a successful product launch (just as there were many failures when not used). For one, a camera manufacturer sent its marketing staff and researchers to a department store to learn about consumer behaviors, questions, and expectations. As a result, a successful family of cameras was launched. In another, a car manufacturer sent its researchers to live in the state where its target customers reside. Some lived with likely customers to learn their experience of using the car. As a result, a successful luxury car was launched and gained a significant market share. The opposite can also be true, especially when a marketer ignores a market that is unlike their own, assuming that everyone is just like them.

Many new products with progressive designs, including kitchen appliances, tools, furniture, car interiors, and electronic gadgets, have used ethnographic research to integrate them into users' lifestyles. The recent focus on design innovation brings new relevance and importance for ethnographic research methods.

SOFTWARE

In the analysis stage of ethnography, there are some very interesting possibilities that are just beginning to emerge. Software platforms include

- 1010Data
- Action
- Amazon Web Services
- Cloudera

- Splunk
- Hortonworks
- IBM
- Infobright
- Cognitio
- Mapr
- Microsoft
- Oracle
- Pivotal
- SAP
- Teradata

The potential of any of these big data analytics engines revolves around the ability to glean knowledge from the way people transact their day-to-day lives. For example, monitoring a Twitter feed in a certain region within a certain ethnic group could expose tremendous understanding of the way these people perceive an issue or product. The potential of this type of analytics in the field of ethnography remains in its infancy.

REFERENCES

Barwise, J. and Perry, J. *Situations and Attitudes*. Cambridge, MA: MIT Press, 1983.

Burgess, R. *In the Field*. London: Routledge, 1984.

Erickson, F. What makes school of ethnography 'ethnographic'? *Anthropology and Education Quarterly*, vol. 15, pp. 51–66, 1984.

Harris, M. and Johnson, O. *Cultural Anthropology*. Needham Heights, MA: Allyn and Bacon, 2000.

LeCompte, M. and Goetz, J. Problems of reliability and validity in ethnographic research. *Review of Educational Research*, Spring, vol. 52, no. 1, pp. 31–60, 1982.

Millen, D.R. *Rapid Ethnography: Time Deepening Strategies for HCI Field Research*. New York: ACM, 2000.

Venkatesh, A. The home of the future: An ethnographic study of new information technologies in the home. *Advances in Consumer Research*, vol. XXVII, Mary Gilly & Joan Meyers-Levy (eds.), Valdosta, Georgia, pp. 88–96, 2001.

von Hippel, E. *The Sources of Innovation*. Oxford, UK: Oscar University Press, 1988.

Wild, H. The making of water cooler logic's stakeholder ethnography—Composting as metaphor for innovation. *Methodological Innovations Online*, vol. 7, no. 1, pp. 46–60, 2012.

11

Generic Creativity Tools

Stephen P. Walsh

CONTENTS

DEFINITION

Generic creativity tools are a set of commonly used tools that are designed to assist individuals and groups in originating new and different thought patterns. They have many common characteristics like thinking positive, not criticizing ideas, thinking out of the box, right-brain thinking, etc. Some of the typical tools are benchmarking, brainstorming, Six Thinking Hats, storyboarding, and TRIZ (theory of inventive problem solving).

USER

The general term *generic creative tools* is so broad that it covers all types of teams and individuals. In the work environment, it particularly focuses on helping teams leverage on the different cultural backgrounds of the individuals that make up the teams to come up with superior solutions.

OFTEN USED IN THE FOLLOWING PHASES OF THE INNOVATIVE PROCESS

The following are the seven phases of the innovative cycle. An X after the phase name indicates that the tool/methodology is used during that specific phase.

- Creation phase X
- Value proposition phase X
- Resourcing phase X
- Documentation phase X
- Production phase X
- Sales/delivery phase X
- Performance analysis phase X

TOOL ACTIVITY BY PHASE

In today's environment, every individual is required to contribute to the organization's success not only physically but also mentally. This means that every individual at every phase of the innovation process has a responsibility to be creative about the way he or she does his or her job, and make suggestions on how the organization can perform better. As a result, many generic creativity tools need to be used effectively by every individual within the organization.

HOW TO USE THE TOOL

Creativity is one of the defining characteristics of humanity. Everyone is creative—despite frequent remonstrations to the contrary by many people! This section is not so much about how to be creative, but rather what stops our innate creativity from being released and so what practical things can be done to remove the blockers?

There are many tools, discussed throughout this book, which will help. At the end of the day, though, that great idea, that awesome inspiration, that beautiful insight will come from *your* mind.

Can creativity be switched on like a light switch? Is it a skill that can be applied according to a timetable? Can organizations manage the generation of ideas? Certainly, it can be switched *off*, or rather many managers (and school teachers or any other profession) have become adept at switching off creativity by the many demotivating practices prevalent in many an office and shop-floor environment.

For sure too, however, there are many examples of *creating to a timetable*—from the wartime examples of the development of Radar or Turing's cracking of the allegedly unbreakable Enigma Code to the daily output of marketing literature and television advertisements. Even Mozart, as an appointed court musician, had to dance to the tune of a prince.

"I'm Not Creative"

The demand for both innovation and for problem solving in all organizations requires the generation of creative thought on a daily basis. So why do so many of us claim to be *not creative*? The issue is one of *mind-sets*—our states of mind, our paradigms that blinker us and deny access to original thoughts. These present themselves as *barriers* or constraints, which we accept without question.

Consider for a moment the following task. You are asked to join all nine dots in the array in Figure 11.1, using just four straight lines, where one line starts from the end of the previous.

With five lines, this would be easy and that is a common response when people are given a task to do with seemingly insufficient resource—"we need more equipment; we need more staff" is a typical outcry to failing to get something done. But in actual fact, the task is do-able if one thinks differently about the way in which one applies the existing resource by challenging the preconceptions of the constraints. The question "why not?" is as powerful as the question "why?"!

In the task of the nine dots, most people when encountering this problem for the first time will restrict themselves to a perceived *box* around the array and do not stray beyond this self-imagined boundary (why not?!). No box was mentioned (though in presenting the example, I did condition the reader to think that way—beware of conditioning!); no conditions were given about the length or width of the lines, only that they be straight and joined end-to-end.

Challenge constraints! Push the boundaries by asking "why not?"—if the constraints are real, they will push back, but otherwise they will dissolve like candyfloss in your mouth.

Figure 11.2 is one solution to the problem, perhaps the origin of the phrase *thinking out of the box.*

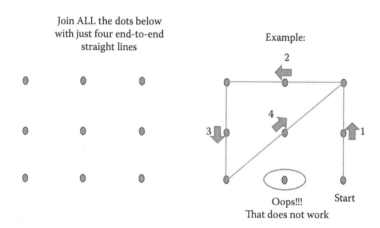

FIGURE 11.1
Nine-point challenge.

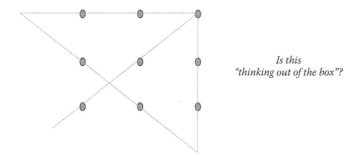

Is this
"thinking out of the box"?

FIGURE 11.2
Answer to the nine-point challenge.

Common Constraints to Creative Thinking

There are many barriers to creativity and they can be categorized as

- Not making the time
- Not having the space (the environment)
- Attitudes of others
- Premature analytical thinking

Let us examine these and illustrate ways in which they might be taken down.

Overcoming the Barriers of Time and Space

Albert Einstein released his creativity often through his famous thought experiments. The day he chose to ride a light beam in his mind's eye was the day our understanding of time, space, and the universe changed! See Figure 11.3.

Many other thinkers have attributed their inspiration to a walk in the woods or along a piece of rugged coastline—effectively giving themselves the space and time to *think*. Allegedly, Newton sat in his mother's Lincolnshire garden when struck by the idea of falling apples, and Archimedes lounged in a hot tub when he had the original *eureka* moment. Einstein's imagineering took his thinking off the conventional mind paths we normally take, allowing new routes to be explored and hitherto unseen doors to be opened.

FIGURE 11.3
Einstein riding a light beam in his mind's eye.

In our daily work, we too need to break the mold of our surroundings. Thinking at the desk is fraught with distractions. Furthermore, our creative output and ability to solve problems can be multiplied greatly if we work as a team—the adage "two heads are better than one" can be improved to "six heads are better than two."

Providing the time and space for teams to meet and address a situation with a view to finding a resolution goes beyond "let's have a meeting to discuss it."

Now a classic approach to generating change, the *Kaizen Blitz* methodology has become a staple of shop-floor and office improvement. Also known as a rapid improvement event (or workshop), the idea behind a short-duration workshop is that a group of people can focus on an issue over just a few days to generate a change plan (or indeed changes). These are "spinning one plate" workshops, as opposed to the traditional way of addressing an issue over many weeks of conventional meetings, and they allow the participants to focus on one agenda item (the issue), without being distracted by the daily noise of operations.

Rapid improvement events are a great way of getting the relevant stakeholders together to share their perspectives, knowledge, and experiences and consequently produce a better quality of ideas for improvement and change. Participants are encouraged to view it as an opportunity to be consulted and involved in making change happen, and typically a facilitator will build an environment of open-mindedness, and positive thinking.

This type of event is necessarily fluid and flexible, though there is an overall process/methodology to follow and it is the responsibility of the facilitator to ensure that the defined outcomes are met. Note that *outcome* is different from *output*: the former will be defined in terms of benefits sought, usually captured as a statement describing the purpose and aims of the workshop, whereas the latter will be the actual actions and results delivered by the participants by the end of the workshop, which should in turn lead to the intended *outcome*.

The resulting quantity and quality of work done in a workshop does truly suggest the existence of the character Doctor Who's TARDIS—workshops appear bigger on the inside than the outside!

To apply this approach, attention to the time and space dimensions is an important factor in producing the aforementioned environment of open-mindedness and positive thinking. There are of course other factors, such as the orchestration of the event by a facilitator, the careful preparation beforehand to clarify aims and objectives, and the application of appropriate tools, such as the ones described in other parts of this literature.

Recent research by scientists at the University of California, Santa Barbara, published in the journal *Psychological Science* in 2012 suggested a strong link between daydreaming and creative thinking. An increasing number of companies have created a literal space for their creative thinking.

Examples that the author has encountered include the technology company E2V, in which there is a specially prepared room where teams go to *think*, called *The Lightbulb*. Royal Mail designed their *iLab—Innovation Lab*—paying careful attention to the color and even shape of the walls (curved not flat surfaces), the seating, and the lighting. They provided playful objects such as guitars and LEGO bricks—not so much to distract but to offer creative rest between creative work. Companies House set aside a valuable meeting room and created a special room for creativity.

Even hiring a hotel conference space for a few days will enable "spinning of one plate" for a group of people, as has been witnessed by the author facilitating many a workshop with local authorities on themes as varied as supplying computers to new starters to services for old people.

When H. James Harrington was chief operating officer of SystemCorp in Montréal, the company had a major problem as they were behind schedule in getting a software product out on schedule for IBM. As a result, they booked rooms in the hotel for three weeks and the team who worked on the product eventually came up with some very creative software approaches. The total team worked 24 hours a day, seven days a week, going to one of the bedrooms to sleep when they got too tired to work. The result was they created a new, very advanced innovative software package that was rated by outside organizations as the most advanced in the field, in a time that would otherwise have been impossible.

The key aspects of managing time and space are as follows:

- Timetable the event and clear participants' diaries for the period of the event.
- Invite participants with plenty of notice.
- Manage…
- Expectations—make sure people are aware of the purpose of the event.

And

- Set up the work area to allow group work—plenty of wall space, natural light, room to move.
- Plan the process but do not predefine the output and design the room layout to suit.
- Apply facilitation tools in the appropriate place (include in your planning of the process).
- Clear your diary of other commitments and distractions.
- Ensure others who are not involved in the event do not disturb you and your participants.

Overcoming the Barriers of Attitude and Premature Analytical Thinking

Attitudes exhibit themselves as behaviors or habits, and habits can be changed. Certainly there is a good proportion of people whose attitude is described as *negative* or *cynical*, and such behaviors would appear to be counterproductive to creativity.

Perhaps they are a consequence of past experiences and failures that reinforce the critical pessimism that so often greets any effort to be creative.

The other day, while shoveling a tonne of manure into my wheelbarrow to transfer it from the front garden where it had been unceremoniously delivered to the rose garden at the back of the house, I came up with an idea for a new design of a wheelbarrow. Figure 11.4 is the sketch I made of it—take a few seconds to look at it—what do think of my design?

On first seeing the sketch, the majority of people say "the wheel's in the wrong place," or even "the tub's too big"; a few offer neutral observations

FIGURE 11.4
Picture of a new wheelbarrow design.

about the use of the barrow, such as "you would have to push it down to move it," and a very small number fail to see how it differs from a traditional wheelbarrow (with the wheel(s) at the front and leg(s) at the back). Hardly anyone—often no one—comments on the potential benefits of the design, namely that it would be more maneuverable, steady when tipped forward and pushing down on the handles to move the heavy-laden barrow would be more advantageous for people with short legs!

Into which camp did you fall? In this exercise, negative comments invariably far outweigh any other comment and, of course, this is a typical first response for many of us—the abominable "no" men (and women) abound. It is a habit. Consider frequently used phrases such as "I hear what you are saying, but..." and "with respect." It is easier to destroy an idea (and with it the originator's motivation and enthusiasm) than to think about possibilities. A positive outlook will require further thinking on our part, while a negative statement ends debate or starts an argument, which equally makes no progress.

Negative attitudes are self-fulfilling prophecies—that new way of doing something or that new design never gets off the ground because people did not put their effort into making it work and they did not *make* the effort because they knew it would not work!

H. James Harrington recommends "that people should be angel advocates rather than devil advocates."

One common behavior, perceived by some as a strength, rather than as a barrier to creativity, is premature evaluation—the logical left-brain thinkers can be a bit too keen to pass early judgment on an idea or proposed problem solution.

Analytical thinking is of course important. It is, however, a later step in the process of innovation. If applied too soon, too little creative thought is expressed.

For example, in brainstorming, one of the key guidelines is given as "no idea is a bad idea," and people recognize that they should not criticize someone's suggestion. This behavior not only demotivates the individual, but it stops the flow of further ideas as people focus on the demerits of the first one to surface.

Similarly, though, positive discussion of someone's idea also stops the flow of further ideas. The aim in this early stage is to proliferate and free-wheel, so both negative and positive discussion should be dissuaded (for now). The time for discussion—analysis—comes later.

Stages of a Plan Activity

What, then, can be done about negative attitudes, preconceptions, and premature analysis? There are three stages of a planned activity, in which attention to certain aspects will minimize the negative impact and turn that negative energy into positive support and contribution:

- Stage 1: Preparation
- Stage 2: Managing the flow—a meta-approach
- Stage 3: Follow-up

Stage 1: Preparation

They say that people do not like change, but is that strictly true? If people were averse to change, they would not move to a new house, apply for another job, or have babies! What does upset people is having change done *at* them—not being consulted about or involved in something that affects them directly.

An important part of preparation for an event is therefore engaging the identified stakeholders in the planning—setting the scene, describing the background and context, consulting about the purpose of a change event (often referred to in the public sector as an *outcomes-based* approach), and establishing the WIIFM factor—the "what's in it for me."

Through discussions with the stakeholders, some of whom will be involved in the upcoming creative event, fears will be voiced and perhaps allayed, expectations will be managed, and above all *clarity* of purpose will be established.

Many negative attitudes will be dealt with by this preworkshop preparation.

Stage 2: Managing the Flow during an Event—A Meta-Approach

An event may be a short meeting, a one-day workshop, or a full-blown three- to five-day rapid improvement event. In all of these cases, the overall approach is the same, what is known as a *meta-approach*, that is to say, the approach of approaches (see Figure 11.5).

This meta-approach is equally applicable for problem solving or decision making and at a higher level for a project or workshop, say, or at a tools level (the affinity diagram is a good example of this, where brainstorming onto sticky notes is followed by arranging the sticky notes into affinity groups with common aspects).

The point of the meta-approach is to uphold a process where action is preceded by analysis, and analysis is preceded by idea generation. In other words, it discourages people jumping to conclusions, acting on preconceptions, or "doing what we have always done."

The approach encourages creativity, the exploration of alternatives, the engagement of stakeholders, and eventually the selection of the best solution (the output of this process) to give the required outcome, with the added bonus of achieving consensus agreement and buy-in.

With a decent facilitator, the approach gives everyone an opportunity to contribute and will quash the objections of the naysayers. The author has witnessed a complete turnaround of negativity with this approach. The approach structures the debate and gives no airing of preconceptions.

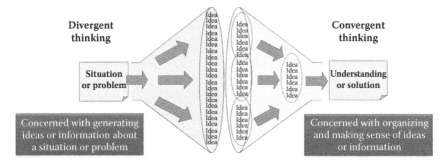

FIGURE 11.5
Meta-approach.

Conclusions are reached through the process of *build* and do not allow ready-made solutions to be presented at the outset.

Figure 11.6 is a breakdown of the generic steps in the meta-approach.

How the approach was able to overcome the barrier of blame was demonstrated in one Kaizen Blitz workshop, where a team of shop-floor operators were grappling with quality issues in their machine shop. They made high-precision fasteners for the aero industry and had narrowed their issues to three characteristics of a bolt—straightness, concentricity, and surface finish.

In their deliberations, they realized that they needed to involve operators from a preform stage before machining, namely people from the in-house foundry who supplied the raw cast bolt.

Imagine the tension in the air when three members of the foundry were invited into the workshop that very afternoon! The potential for a verbal punch-up was high, but the facilitator (this author) introduced the group to a simple fishbone diagram technique to capture all the variables for each of the bolt attributes and they generated three fishbones, one for each characteristic. By simply asking the group to consider the question "what aspects of the bolt affect the straightness?," the *concentricity* and then *surface finish*, all minds turned positively to brain dumping what they understood about the process, and arguments about whose fault it was that such-and-such occurred were circumnavigated.

The three fishbones were amalgamated into a matrix of output characteristics versus variables (known as a Y2X matrix), and major contenders for investigation were identified. A collaborative plan was then drawn up to conduct data collection and analysis of the major potential causes of issues with straightness, concentricity, and surface finish and a study conducted. No fights, no ill will, no problem!

The approach, using the tools of fishbone diagrams (divergent thinking) and the Y2X matrix (convergent thinking), demonstrated very well how negative attitudes, domineering personalities, and jumping to conclusions based on preconceived ideas could be avoided. The author later discovered that this particular issue and feud had been going on for several years!

This meta-approach can be seen in many established improvement methodologies, for example, in the Ford 8D problem-solving process or the Six Sigma DMAICT methodology.

Figure 11.7 shows de Bono's *Six Thinking Hats* (described elsewhere in this book), and Figure 11.8 shows how the meta-approach can be overlaid onto the de Bono process.

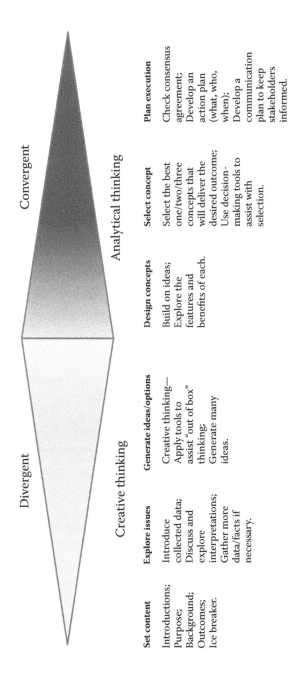

FIGURE 11.6

Breakdown of the generic steps in the meta-approach.

de Bono's Thinking Hats:

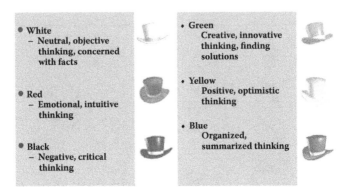

Process for using the hats

State an issue (e.g., a current issue that the team is dealing with)

Step 1 (White) — Present facts of the case

Step 2 (Green) — Generate ideas of how to handle

Step 3 (Red) — Gut feelings

Step 4 (Yellow) — Evaluation the benefits

Step 5 (Black) — Evaluate the concerns

Step 6 (Blue) — Summarize

FIGURE 11.7
Six Thinking Hats.

Divergent

Convergent

Creative thinking

Analytical thinking

FIGURE 11.8
How the meta approach can be overlaid onto the Six Thinking Hats process.

Walt Disney, too, engaged the meta-process for his (and his team's) generation of innovative, practical ideas. It is said that he had three rooms, each containing an armchair in which he sat, one after the other (witness again the importance of making *time* and *space* for idea generation).

In the first, he allowed himself to *dream*—allowing his mind to wander freely with no boundaries or restraints; in the second—his *realist* chair—he would organize the ideas he dreamt of: What would be needed to make them real? How could it be done? He would consider practical plans to convert the dreams into reality. Finally, his third chair represented the *critic*. Here, he would look for the difficulties and unintended consequences, asking "What could go wrong?," "What's missing?," and then evaluate the plans, turning them into action.

Stage 3: Follow-Up

The third stage is not about creativity, but is essential if the creative thinking generated in an event is to be harnessed practically as a solution to a problem or as an action that produces results. The follow-up has to be planned (begun during the event) and managed. This stage is one for the completer–finishers of the group—the *i* dotters and *t* crossers, but everyone is involved to turn *dreams* into *reality* (to refer back to Disney's creativity strategy).

Concept designs (for a process or a piece of hardware) need to become detailed designs, costed, refined, and built. Decisions need to be communicated, promoted, and executed. Resulting changes need to be documented, monitored, and reported.

Helping the Creative Thinking Process

Attention to the aspects of time and space and using the divergent–convergent meta-approach to plan and structure an event will enable creative thinking to be released. Two other factors will help smooth the flow, namely the appropriate selection and application of so-called creativity tools and the use of a facilitator.

Role of a Facilitator

A facilitator's role is to manage the flow of an event in order to achieve the stated outcome. He or she should be impartial and objective, and therefore what the *outputs* are is less important to them as that there *are* outputs delivered that are satisfactory, that is to say that they will deliver the *outcome*. This is a purest view and some divergence from this is sometimes accommodated.

To manage the flow, the facilitator will most normally be part of the preparation stage described above, including the planning of the event. Part of that will be deciding what tools to use and so the facilitator typically has a kitbag of tools, upon which he or she can call.

The other aspect of managing the flow is monitoring progress toward the achievement of the objectives and recognizing any impedance, such as any lack of understanding, confusion, or misbehavior. A good facilitator will know how to evade these things before they are problematic by using techniques of scene setting, regular reviews, interpersonal skills, and the timely use of divergent–convergent tools.

The outcome of that facilitation is an environment in which the knowledge and experience of the participants can be shared, and their engagement and creativity can flourish.

Role of Creativity Tools

Tools help to smooth the flow of the process in the preparation stage and in the divergent and the convergent phases of an event. Indeed, the tools described in this book can be mapped onto the chronology of an event, and this represents a key part of planning the event itself. Some tools are naturally applicable to the divergent phase (e.g., brainstorming), while others are suited to convergent, analytical thinking (any of the decision-making tools). Some tools comprise both elements—the affinity diagram is one such tool, as mentioned earlier.

The creativity tools are described in other parts of this book, and so their descriptions will be omitted here. It is enough to say that there are generic categories for the tools:

- Brainstorming derivatives
- Association techniques
- Benchmarking

Brainstorming derivatives include reverse brainstorming, brainwriting, synectics, affinity diagrams, and fishbone diagrams, to name a few.

Even mind mapping, as popularized by the great memory man *Tony Buzan*, can be regarded as a personal brainstorm (a group mind map is sometimes called a spray diagram).

They all operate on the principle of a thought cascade, where one thought will catalyze another and another, each thought spraying out more thoughts until there is a veritable cascade of ideas released, rather like a nuclear chain reaction releasing energy into the surroundings in abundance. That is why some proposed the renaming of *brainstorm* to *thought shower*.

To proliferate, the cascade has to be unimpeded by analysis, discussion, or criticism with the focus on getting a maximum number of ideas in a short space of time. Hence, the usual rules of brainstorming apply and anything goes.

Association techniques work on the principle of *out-of-box* thinking and encourage the participants to discover avenues of thought that they may not have considered without a little inspiring dreaming.

One of the barriers of the human mind is mind grooving. We have a tendency to routinize our approaches to tasks—after all, this simplifies our lives, makes us efficient at the task, and so allows us to complete the task as quickly as possible. We develop a mind-set—which is no bad thing until we need to change that mind-set.

For example, recall for a moment your early days of learning to drive a car (or a bicycle, if you are a nondriver). Much of your mind was occupied thinking about the mechanics of making the machine move, steering and braking, and less so about sharing the road with other users. Years later, driving or cycling has become a subconscious competence and, presumably, your attention can be given to the important task of driving safely with others.

But what happens when you hire a car where the traffic drives on the opposite side of the road? How long is it before you stop bruising your elbow on the door panel or looking in the wrong place for the rearview mirror? (The same principle applies to cyclists—some bicycles brake by reverse-pedaling—a fact the author discovered rather late while negotiating a busy roundabout—which had a contrary flow, to boot).

Our mind grooves become established paths, which we frequent and reinforce, and like the little creature that finds a safe and quick route to the waterhole, we wear a path that becomes difficult to change.

The trouble is, we are not aware that we are riding in a groove, and this limits our capacity to think laterally.

Association tools work on the principle of encouraging us to take our mind on a path that seemingly has nothing to do with the issue on the table. You may be asked, for example, to imagine and verbalize a walk up a mountain, or to take a literal walk in the grounds and bring back five objects you find.

Comparisons are then made between the issue and the story or objects, the aim being to open up doors in the mind that were hitherto invisible. In essence, association techniques boost the mind out of its groove.

Common and well-documented association tools include

- Heroes and villains: *What would Batman do in this situation? Or Genghis Khan?*
- Similarities and differences: *Comparing an item from nature, say, with the issue (compare a natural process/system with a designed process/ system or a natural "thing" with a designed "thing").*

- Story telling: *the "walk up the mountain."*
- SCAMPER: *a substitution approach.*

Benchmarking is a methodology in itself and has countless literature written about it. The premise is one of learning from others by comparing one's own performance or product with that of another organization (or department in a large organization). This can be a source of inspiration, and when witnessing how some other organization achieves a common outcome, it can be a positive challenge to one's own mind-set, resulting in some creative ideas.

Many think of benchmarking as an activity of comparing oneself with the competition, but it is with any organization (though usually with someone who can demonstrate outstanding performance). Focus is important, so selecting a specific area is more fruitful than what some call *industrial tourism*.

The process of benchmarking is to

- First understand the drivers—what do you need to achieve, where are you failing?
- Then understand your process (or product) relating to the drivers.
- Select a partner with whom to benchmark (and remember it is a sharing exercise).
- Conduct the study.
- Examine the collected data/evidence.
- Consider changes to your own process (adopt/adapt the learning— be creative).
- Make the change and monitor the result.

Note again the allusion to the meta-process of divergence–convergence.

List of Tools

In the book *The Creative Toolkit—Provoking Creativity in Individuals and Organizations* published by McGraw-Hill in 1998, the authors listed a generic creativity tool summary. The headings in this tool summary are as follows:

Mind Expanders
- The 2-minute mind
- Mindbeats
- The alphabet

- The numbers
- A nursery rhyme
- Common objects
- Personal creativity
- Analyzing outrageous ideas
- Pictures to drive creativity
- Words to drive creativity
- Differences and similarities
- Defining other applications
- Creative progress reports
- Dreaming in color
- Recording your evening's activities
- Discarding the boom box

Creative Tools

- Brainstorming
- Card sort/affinity diagram
- Cards and games
- Cartoon drawings
- Code talk
- Conversations
- Drawings
- Environment
- Exaggerated objects
- Five whys
- Force analysis
- Free association
- Games of chance
- Listening for comprehension
- Manager to manager event
- Measles charts
- Mental games
- Mind mapping
- Must/wants/wows
- Partitioning
- Physical games
- Physical models
- Pick-up sticks
- Polar cases
- Possibility genders

- Presentations
- Rainbow flowcharting
- Reality matrix
- Role-playing
- Say/think
- Selection window
- Service ladder
- Skill games
- Song titles
- Thinking words
- Tours

Now, although these tools and techniques may sound general in nature, they all have a step-by-step procedure behind them, providing directions on how to use them effectively.

SUMMARY

Creativity is in us all. There are, however, many barriers that cause that creative spark to dim. This chapter has proposed ways in which the barriers of time, space, attitude, and preconceptions can be overcome.

A few straightforward guidelines have been suggested to deal with them—understanding what you are trying to achieve, preparation of the environment, and the expectations of participants and managing the flow of a creativity event through good facilitation and appropriate use of tools against the backdrop of a divergent–convergent meta-process have been described.

The author's own experience has demonstrated that all groups can share and combine their knowledge and experience to produce something that is more than the sum of the parts.

Something new has been created.

EXAMPLES

See the examples included in the section entitled "How to Use the Tool."

SOFTWARE

Some commercial software available includes but is not limited to

- MindMap
- Smartdraw
- QI macros: http://www.qimacros.com

SUGGESTED ADDITIONAL READING

Asaka, T. and Ozeki, K., eds. *Handbook of Quality Tools: The Japanese Approach.* Portland, OR: Productivity Press, 1998.

Harrington, H.J. and Lomax, K. *Performance Improvement Methods.* New York: McGraw-Hill, 2000.

12

HU Diagrams (Harmful–Useful Diagrams)

H. James Harrington

CONTENTS

You optimize the system by eliminating all the harmful effects.

H. James Harrington

DEFINITION

Harmful–useful (HU) diagrams are an effective way of providing a visual picture of the interface between the harmful and useful characteristics of a system or process.

USER

This tool can be used by individuals, but its best use is with a group of four to eight people. Cross-functional teams usually yield the best results from this activity.

OFTEN USED IN THE FOLLOWING PHASES OF THE INNOVATIVE PROCESS

The following are the seven phases of the innovative cycle. An X after the phase name indicates that the tool/methodology can effectively be used during that phase.

- Creation phase X
- Value proposition phase
- Resourcing phase
- Documentation phase
- Production phase X
- Sales/delivery phase
- Performance analysis phase

TOOL ACTIVITY BY PHASE

- Creation phase—This tool is used during this phase to evaluate the excellence of a proposed solution or to help define the next evolutionary step in a product development cycle.
- Production phase—The HU diagram can be used to evaluate alternate production approaches and to solve production-related problems.

HOW TO USE THE TOOL

HU diagrams are new approach to defining how to correct problems or to improve designs related to product and process. It also focuses on risk management to ensure the solution has minimized the potential risks related to the solution. Sometimes, HU diagrams are called *contradiction diagrams*. They get better results than the old brainstorming approach.

Information about HU Diagrams

To date, the tools used by performance improvement professionals have been mostly directed at process analysis tools, such as *ask why five times*. These tools have mostly been directed at defining and identifying root causes of problems. Now, at last, we have a problem-correcting tool called HU diagrams, which is sometimes called contradiction diagrams. This approach is designed to guide the user to an effective resolution of a problem. In today's environment, we need to do more than just think outside of the box. We need to tear down all the walls of the many boxes that we compartmentalize ourselves in (see Figure 12.1) (Harrington et al., 2011).

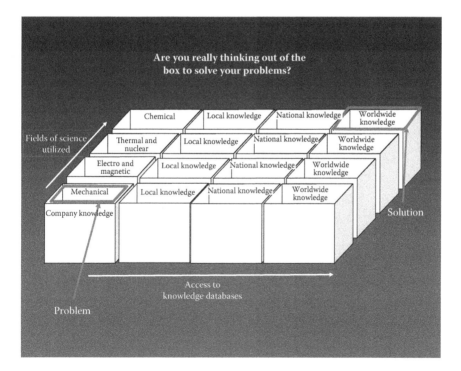

FIGURE 12.1
Boxes we build around ourselves.

All processes have positive aspects that we call useful functions, and all processes also have negative aspects called harmful functions; these are represented in the HU diagrams by the symbols in Figure 12.2.

HU diagrams only use two symbols—harmful functions (undesirable features) and useful functions (desirable features). The arrows that connect them can be designated as one of two relationships, as shown in Figure 12.2.

The arrow from one symbol to the other symbol indicates that the first symbol established the relationship to the other symbol. The arrow without a vertical line through it indicates that the first symbol caused or produced the other symbol to exist. The arrow with the vertical line through it indicates that the first symbol counteracts or inhibits the second symbol. Sometimes, a useful function can cause another function that is desirable to occur. However, sometimes a useful function has undesirable side effects and causes something harmful to happen. On the other hand, a harmful function can cause a harmful function to occur or it could cause a useful function to occur (see Figure 12.3). An HU diagram is essentially a collection of cause-and-effect relationships describing various situations.

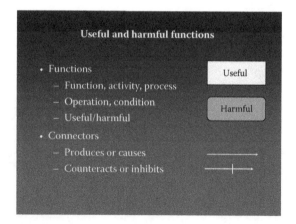

FIGURE 12.2
Useful and harmful symbols and the two types of connectors.

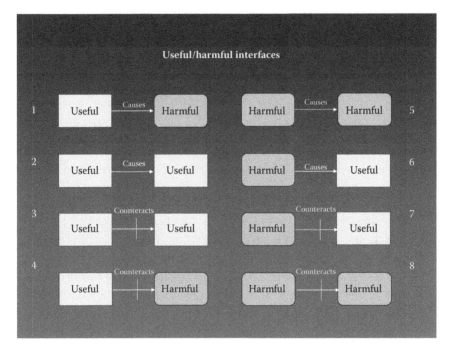

FIGURE 12.3
Two functions can be related in eight different ways.

Contradictions

A contradiction is when something useful has undesirable side effects and causes something harmful to happen. It can also apply when something harmful has desirable side effects and causes something useful to happen.

Contradictions are the undesirable situations that fall into four categories. Relationships 1, 3, 6, and 7 in Figure 12.3 are all contradictions as they are two opposite functions connected together. The situations when two opposite functions are connected are not necessarily contradictions. It also depends on the time of the connector.

Three Types of Contradictions

- Type 1 category is when a function produces a similar function but also produces an opposite function. For example, a useful function produces another useful function (desirable) but also produces a harmful function (undesirable) (see Figure 12.4).

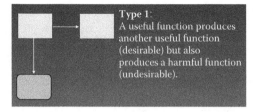

Type 1:
A useful function produces another useful function (desirable) but also produces a harmful function (undesirable).

FIGURE 12.4
Useful function that produces another useful function and a harmful function (type 1 relationship).

- Type 2 category is when a function counteracts an opposite function but also produces another opposite function. For example, a useful function counteracts a harmful function (desirable) but produces a harmful function (undesirable) (see Figure 12.5).
- Type 3 category is when a function counteracts an opposite function but also counteracts a similar function. For example, a useful function counteracts a harmful function (desirable) but also counteracts another useful function (undesirable) (see Figure 12.5).

A process without contradictions would be the ideal process, but in reality, there is no such thing as a completely ideal process.

All processes have at least one contradiction. In fact, the reason for analyzing a process is to maximize the useful elements and minimize the harmful elements—in other words, to maximize the real value-added content while minimizing the non-value-added content of the process.

Basically, there are four ways to resolve contradictions:

- Separation in space—find a space in which the harmful function does not occur.
- Separation in time—perform the useful function at a time when the harmful effect does not occur.
- Separation in structure—find a structural level at which the harmful function does not occur.
- Separate in conditions—find a condition under which the harmful function does not occur.

Let us look at one of the four options—separation in time. If you were to focus on separation in time, the following things should be considered:

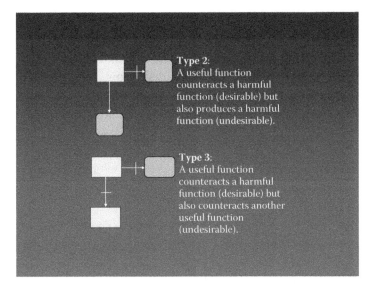

FIGURE 12.5
Example of useful functions that produce type 2 and type 3 relationships.

- Separate opposite requirements in time
- Preliminary action
- Partial preliminary action
- Preliminary placement of an object
- Create and use pauses
- Staggered processing
- Dynamization
- Use postprocessing time
- Concentrate energy
- Predestruction
- Reduce strength
- Preliminary stress
- Poly-system with shifted characteristics
- Decrease stability
- Transition from stationary to mobile
- Divide into mobile parts
- Apply physical effects
- Add a mobile object
- Use interchangeable objects
- Use elements with dynamic features
- Use adjustable elements and links

To get a better understanding of the harmful and useful functions, let us consider a lawn mower. The primary function would be to cut grass, which is a useful function. To move that one step lower, think about what are the primary functions that allow it to cut grass. These would include blade rotation, blade has sharp edges, mower has wheels, etc. Each of these is a useful function. For each of these useful functions, we can expand by saying what causes a useful function to operate. For example, blade rotation is driven by an engine that turns the shaft; wheels move based on a person pushing them—each of these is a useful function and they enlarge on the previous useful function. Continuing with this line of reason, what causes the engine to turn? What resources or raw materials are needed? In this case, internal combustion, gasoline, a gas tank, air, air intake, electrical spark, spark plugs—each of these is a useful function or factor.

Now let us change our stream of thought to some of the negative things, for example:

- Throwing objects
- Cutting feet and hands
- Toxic exhaust
- Engine heat
- Gasoline cost
- Clippings clog motor
- Motor noise
- Pushing is tiring

Each of these candidates is a harmful function. When you identify a function, ask yourself if there is any side effect or by-product. For example, the internal combustion side effects are heat, motion, component wear, noise, exhaust, etc. Whenever you identify a harmful function, ask yourself what caused it or what else has to function for it to happen. For example, for internal combustion you need

- Air
- Kill switch closed
- Spark plugs
- Electricity

All of these could be useful or harmful functions depending on the situation. When you identify a function, ask yourself what inhibits this from happening. For example, for internal combustion, we have

- The ignition switch
- Dirt in gas
- Broken wire
- Water in air
- Clogged blade
- Etc.

We will now look at a problem related to an airplane's jet engine containment ring (see Figure 12.6).

Figure 12.7 is a simple HU diagram of a problem with the fan in a jet engine breaking and causing damage to the airplane. The fan rotates at high speeds (center useful symbol), causing a large quantity of air to move through the engine (useful symbol to the left of center). This is a useful function as it causes the plane to move. The fan rotating at high speeds causes two detrimental or harmful things to occur: the centrifugal force applies high pressure to the impellers (blades) on the fan, and this high pressure can cause the impellers to burst. The particles from the impellers can damage the airplane. At the same time, the high-speed rotation of the fan gives high energy to the fragments that are flying, which can cause damage to the airplane.

Figure 12.8 is a more complete HU diagram of the problem with the fan in a jet engine breaking and causing damage to the plane.

In general, each box on this picture represents an opportunity to address the problem situation. The four circled areas in Figure 12.9 indicate areas that must be addressed first to offset the harmful parts of the diagram and bring into better balance the ratio of harmful and useful functions.

Information about an Operator

An operator is a little nugget of wisdom (recommendation, suggestion) that may be used to change the system design. They trigger you into thinking how to solve the problem or to improve the process under evaluation. Operators are drawn from successful results of previous action that

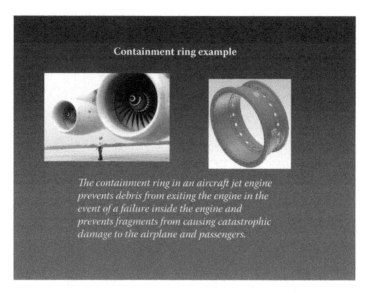

FIGURE 12.6
Jet engine containment ring problem.

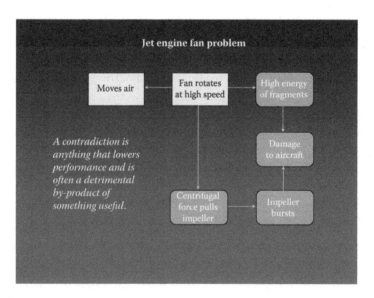

FIGURE 12.7
Starting-point HU diagram of the jet engine problem.

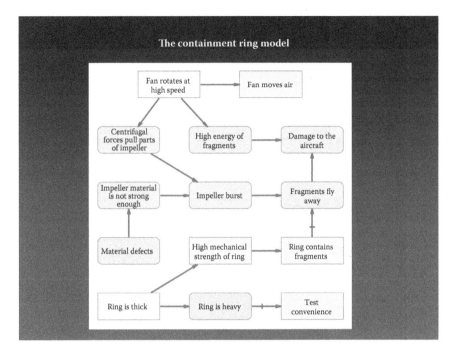

FIGURE 12.8
Jet engine fan problem detailed HU diagram.

resolve different technology problems or process problems. To date, there are approximately 1000 operator principles that have been defined. It is our experience that between 200 and 500 of these principles are all that are really needed to solve most problems a process improvement team (PIT) will encounter.

The shaded area that includes the useful symbol labeled *Test convenience* and the harmful symbol labeled *Ring is heavy* are areas that present improvement opportunities to reduce the weight of the containment ring. In this case, the PIT will use an operating principle called *Uniformity*— changing an object's structure from uniform to nonuniform, changing an external environment (or external influence) from uniform to nonuniform. Using this operator as a starting point, the PIT will adapt it to the conditions set up in the shaded area. A typical idea that could come out from this analysis would be to change the ring thickness (kind of like a two-liter soft-drink bottle) over its length and over its width, making the ring denser and closer in to the blades and directly in line with the blade's motion, but less dense everywhere else.

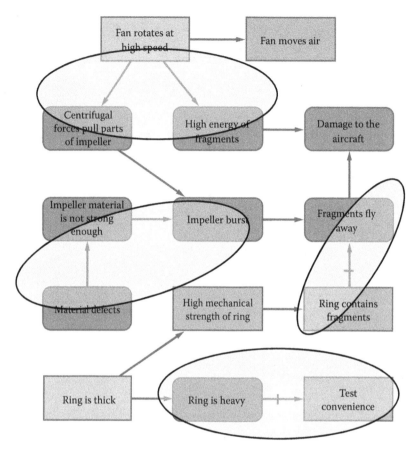

FIGURE 12.9
Jet engine fan problem detailed HU diagram with four areas identified.

Another operator principle that could be applied is asymmetry—changing the shape of an object from symmetrical to asymmetrical. This leads the PIT to suggest that the side of the ring closest to the body of the airplane could be thicker to give maximum protection. There would be less damage if the fragments hit the side away from the body of the plane.

The shaded area that includes the useful symbol labeled *Ring contains fragments* and the harmful symbol labeled *Fragments fly away* presents another opportunity for improvement. In this case, the PIT will use an operating principle called *porous materials*—making an object porous or adding porous elements (inserting insert, coatings, etc.). Using this principle, the PIT could suggest that a honeycomb structure would be more effective. The sharp edges of the structure would act like knives to shred the fragments.

The shaded area that includes the harmful symbol labeled *High energy of fragments* and the harmful symbol labeled *Damage to the aircraft* also presents an opportunity for improvement. In this case, the PIT will focus on reducing the energy of the fragments. To accomplish this, the PIT will use two operating principles: adding porous materials makes an object porous, or adding porous elements (inserts, coatings, etc.) and composite materials changes uniform materials to composite (multiple) materials. On the basis of using the concepts in these two operators, the PIT could suggest using multiple lightweight rings. The first is thin and stiff but porous. This will shred the blade fragments but does not stop them. The second ring is made from carbon fiber, which is strong and lightweight and stops the fragments.

There is a fourth opportunity available even though it includes two useful symbols. It is the shaded area that includes the symbol *Ring is thick* and the useful symbol called *High mechanical strength of ring*. In this case, the PIT could use an operating principle called the *nested doll*—place one object inside another; place each object, in turn, inside the other. By applying this operator to the situation, the PIT could come up with an idea of using two containment rings, one inside the other. The inside ring could be thicker than the outside one, but put together and it will be lighter than one big ring.

The example we used was a design solution, but HU diagrams work equally well for process solutions. Figure 12.10 is an HU diagram for the impact of change on an organization. Figure 12.11 is an HU diagram of a mass-market problem highlighting two areas where additional actions need to be taken to offset the harmful effects.

Figure 12.12 shows how a more in-depth study that can be taken related to each of the two target areas using additional submaps of the HU diagram.

Figure 12.13 is a typical HU diagram generated by a computer program along with the comments related to key elements on the diagram.

SUMMARY

We have found HU diagrams to be very effective in helping PITs generate very innovative solutions, particularly when they are combined with a knowledge-based system like I-TRIZ. The I-TRIZ system contains a world

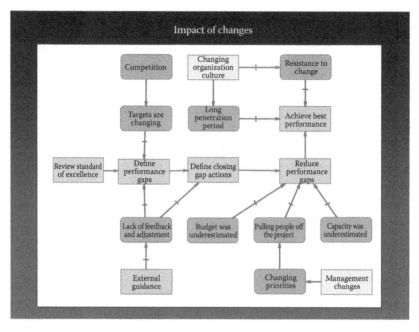

FIGURE 12.10
HU diagram of impact of change on an orginization

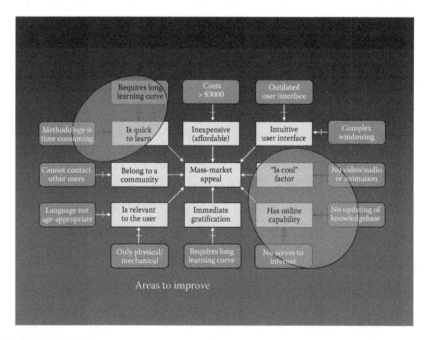

FIGURE 12.11
HU diagram of mass-market problem. Identifying two improvement areas.

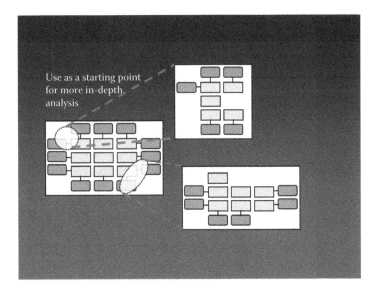

FIGURE 12.12
HU diagram of mass-market problem.

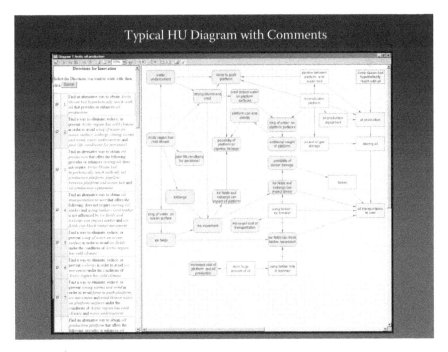

FIGURE 12.13
Typical HU diagram with comments.

of experience from problems that have been solved in the past and whose approaches continuously repeat themselves in being useful in solving future problems. It is an effective approach for defining the operator that relates to the problem the PIT is addressing with practical examples of how to use each of the operators.

I-TRIZ or Operating System for Innovation is a software supported system sold by 20–20 Innovation and Ideation Inc. It is made up of four software-supported packages (see Figure 12.14) (Zlotin and Zusman, 2001).

I-TRIZ packages use HU diagrams as their analysis tool. They construct the HU diagram using their computer programs. We suggest that you start with the Inventive Problem Solving software package, although the Directed Evolution software package is a better but more complex methodology. Once the future-state solution is defined, we suggest you use the Failure Prediction software package to define the major risks related to the new process.

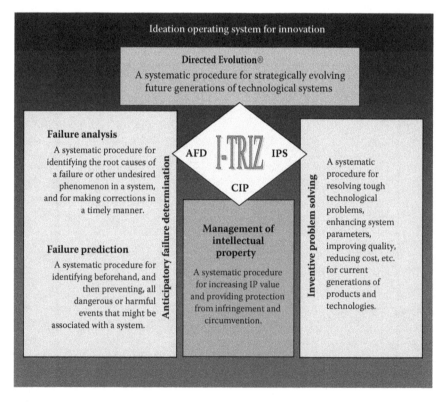

FIGURE 12.14
Four software packages that make-up I-TRIZ.

EXAMPLES

See the examples that were embedded in the section of this chapter entitled "How to Use this Tool."

SOFTWARE

- Anticipatory Failure Determination (AFD)—Ideation: www.idea tiontriz.com
- Directed Evolution® (DE)—Ideation: www.ideationtriz.com
- Inventive Problem Solving (IPS)—Ideation: www.ideationtriz.com

REFERENCES

Harrington, H.J., Fulbright, R., and Zusman, A. HU goes there. *Quality Progress*, September 2011.

Zlotin, B. and Zusman, A. *Directed Evolution: Philosophy, Theory and Practice.* Southfield, MI: Ideation International Inc., 2001.

SUGGESTED ADDITIONAL READING

Altshuller, G. *Creativity as an Exact Science.* New York: Gordon and Breach Science Publishers Inc., 1984.

Ideation International Inc. *Tools of Classic TRIZ.* Southfield, MI: Ideation International Inc.,1999.

Terninko, J., Zusman, A., and Zlotin, B. *Systematic Innovation: An Introduction to TRIZ (Theory of Inventing Problem Solving).* Boca Raton, FL: St. Lucie Press, 1998.

Visnepolschi, S. *How to Deal with Failures (The Smart Way).* Southfield, MI: Ideation International Inc., 2008.

Zlotin, B., Zusman, A., and Fulbright, R. Knowledge based tools for software supported innovation and problem solving. Presented at TRIZCON, 2011.

13

I-TRIZ (Ideation TRIZ)

Alla Zusman and Larry R. Smith

CONTENTS

DEFINITION

I-TRIZ is an abbreviation for Ideation TRIZ, which is a restructuring and enhancement of classical TRIZ methodology based on modern research and practice. It is a guided set of step-by-step questions and instructions that aid teams in approaching, thinking, and dealing with systems targeted for innovation. It provides specific practical team guidance for the following applications:

- Solving a nontechnical or business issue
- Solving a technical or engineering issue
- Finding the root cause(s) of a system issue
- Anticipating and preventing possible system issues
- Predicting and inventing specific innovative products or services customers will want in the future
- Patent (invention) evaluation, disclosure preparation, and enhancement to either work-around (invent around, design around) an existing (blocking) patent or provide a patent *fence* to prevent possible work-arounds.

USER

I-TRIZ can be used by an individual who needs some quick, out-of-the-box ideas. However, this methodology is most effective when used with a cross-functional team composed of people with different knowledge and perspectives of the system and issues.

OFTEN USED IN THE FOLLOWING PHASES OF THE INNOVATIVE PROCESS

The following are the seven phases of the innovative cycle. An X after the phase name indicates that the tool/methodology is used during that specific phase.

- Creation phase X
- Value proposition phase X
- Resourcing phase X
- Documentation phase X
- Production phase X
- Sales/delivery phase X
- Performance analysis phase X

TOOL ACTIVITY BY PHASE

- Creation phase—All of the I-TRIZ features described above enable idea generation for the creation phase.
- Value proposition phase—Both TRIZ and I-TRIZ are specifically focused to define and improve the value proposition. The particular activities involve:
  - Understanding and applying appropriate win/win principles to existing system conflicts to improve customer satisfaction and reduce waste.
  - Identification and application of existing resources that exist in the system, supersystem, and subsystems. Use of resources improves value by increasing functionality while reducing cost.
  - Understanding and improving ideality: As defined by Boris Zlotin (Ideation International Inc., 2005), ideality is the sum of all the useful functions (features and benefits) over the sum of

all the harmful functions (costs, undesired side effects, etc.) of a system. A focus on improving ideality enhances value.

- Resourcing phase—Resources associated with financing are identified for team use. Specific conflicts associated with financing can also be modeled and addressed with I-TRIZ.
- Documentation phase—The guided step-by-step questions and instructions, the system model(s), directions for innovation, idea generation utilizing systems of operators, and the organization of these ideas into prioritized concepts are completely documented for team use. I-TRIZ can also be used for invention disclosure preparation.
- Production phase—I-TRIZ can be used to fix, anticipate, and prevent issues in the production phase.
- Sales/delivery phase—I-TRIZ can be used to fix, anticipate, and prevent issues associated with sales and delivery.
- Performance analysis phase—I-TRIZ can be used to assess the following:
  - Existing utilization of laws and lines of evolution associated with current technology and performance
  - Degree of ideality
  - Use of available resources
  - Existing functional conflicts in the system
  - Future developments due to evolution of the system

HOW TO USE THE TOOL

Introduction

The following instruction in selected I-TRIZ elements is by no means comprehensive. It is intended to enable a beginner to start using introductory I-TRIZ thought processes and selective tools while preparing the reader for more advanced work.

I-TRIZ: Ideation TRIZ (where TRIZ is a Russian acronym for *theory of inventive problem solving*, described in Chapter 29) is a restructuring and enhancement of classical TRIZ methodology based on the modern research and practice of Boris Zlotin, Alla Zusman, Svetlana Visnepolschi, Vladimir Proseanic, and others at Ideation International Inc. On the basis of the study of more than 3 million patents and practical application in thousands of actual case studies, I-TRIZ features the following:

- A guided set of step-by-step questions and instructions that aid teams in approaching, thinking, and dealing with systems targeted for innovation
- A method for modeling the interaction of system elements (Harrington et al., 2012) to identify directions for innovation in system areas that provide the greatest leverage for change
- An enhancement, expansion, and restructuring of the TRIZ body of knowledge to more effectively and efficiently guide team brainstorming by utilizing systems of operators organized by laws (patterns) and lines of system evolution. The I-TRIZ body of knowledge consists of
  - Approximately 440 operators for technical applications
  - Approximately 140 operators for nontechnical applications
  - Twelve laws/patterns and more than 600 lines of system evolution that detail how systems evolve over time
- Specific practical team guidance for the following applications requiring innovation:
  - Solving a nontechnical or business issue
  - Solving a technical or engineering issue
  - Finding the root cause(s) of a system issue
  - Anticipating and preventing possible system issues
  - Predicting and inventing specific innovative products or services customers will want in the future
  - Patent (invention) evaluation, disclosure preparation, and enhancement to either work-around (invent around, design around) an existing (blocking) patent or provide a patent *fence* to prevent possible work-arounds

Innovative Situation Questionnaire®

- Innovation Situation Questionnaire (ISQ): In their practice, Zlotin and his colleagues (Ideation International Inc., 2005) found that a number of teams were creating and implementing solutions that had little or no impact on the original issue they wanted to address. Obviously, the teams really did not understand what they needed to accomplish. An ISQ is a set of questions that, when considered by a team, will help the team better understand their mission. In the American Society for Quality (ASQ) Education and Training and Initiative (ETI) case study, the team originally thought their purpose was to increase revenue. After considering the questions in the ISQ (see the ASQ Example case study in this chapter), they understood that their task was "The development and delivery of educational events involving quality technology, concepts, and tools to enable members to improve themselves and their world." There are several types of ISQs depending on the I-TRIZ application. All ISQs document the following:

- Problem description
- System information (structure, functioning, primary useful function, reason to perform the primary useful function, environment)
- Problem situation (problem mechanism, consequences, history, other systems with the same problems, other problems in the system)
- Ideal vision of solution
- Resources
- Allowable changes to the system
- Criteria for selecting solution concepts

Understand and Mobilize Resources

The number of possible innovative ideas is directly correlated to available resources. Teams are encouraged to consider and document the following resources that are available in the system, supersystem, and subsystems:

- Field resources: All types of energy (mechanical, thermal, chemical, electrical, magnetic, electromagnetic), action, or force
- Functional resources: The opportunity for the system, supersystem, or subsystem to perform additional functions, including supereffects (unexpected benefits that occur as a result of innovation)
- Information resources: Knowledge about the process, system elements, fields, system functions, and additional information about the system (which can be obtained with the help of dissipation fields, or matter or fields passing through the system)
- Substance resources: Parts, components, materials used to compose and to operate the system, supersystem, or subsystem (including by-products, waste, etc.)

- Space resources: Free or unoccupied space, voids, or new space created by rearrangement of system elements or through a change in form
- Time resources: Activity preparation time, duration of work operation, breaks, idle time, time after process completion; time intervals before the start, after the finish, and between the cycles of a technological process, which are partially or completely unused
- Derived resources: Derive new resources from combinations of the above

System Approach

By considering the system, subsystem, and supersystem inputs/outputs, causes/effects, past/present/future, as illustrated in Figure 13.1; the team considers how change in any of these aspects could be utilized to improve the situation.

Identify Patterns Associated with System Conflicts

Look for pattern(s) in the system similar to Figure 13.2, where there is a conflict between two or more system functions (where the improvement in one function of a system causes another to degrade), and behind this conflict is a function that is conflicting with itself (the function *wants* to be in two states at the same time, or wants to exist or not exist at the same time). Examples are shown in the ASQ case study in this chapter.

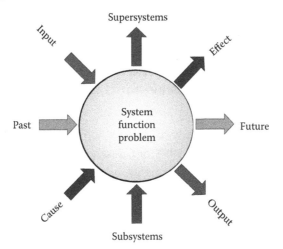

FIGURE 13.1
Ideation's system approach to idea generation.

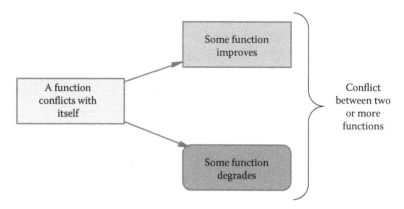

FIGURE 13.2
System conflict patterns that lead to directions for innovation.

Select Direction(s) for Innovation

Utilizing system pattern(s) similar to Figure 13.2, select one or more of the following directions for innovation:

- Resolve the contradiction associated with a function conflicting with itself.
- Counteract a harmful function (address the function that is degrading the system).
- Improve a useful function (make the function that is improving the system even more useful, or find an alternative to this function free of the current conflict).

Utilize Appropriate Groups from the System of Operators to Facilitate Brainstorming

Systems of operators (Figure 13.3) are used to guide team brainstorming. These systems of operators utilize principles in Altshuller's original matrix (see Terninko et al., 1998, for descriptions and examples of the principles) with additional principles, all organized in a more efficient and effective way. Figure 13.3 is a series of nested I-TRIZ operators that are not comprehensive but will get the reader started in effective systematic brainstorming.

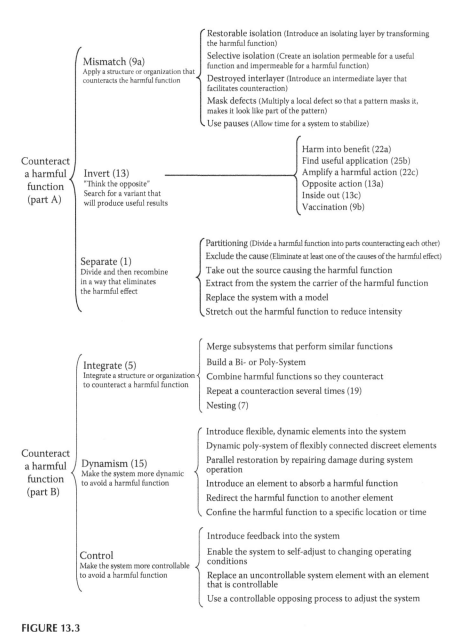

FIGURE 13.3

Series of nested I-TRIZ operators.   (*Continued*)

Improve a useful function (part A)

Match (3c)
Match a particular structure of the system to optimize the overall performance of the system

- Replace a universal system with a set of specialized systems
- Enable a system to perform more useful functions
- Sort elements into groups and treat each group separately
- Equi-potentiality: create a process or super-system that removes barriers
- Synchronize simultaneous processes
- To perform two incompatible functions, perform one function in pauses of the other

Invert (13)
"Think the opposite" search for a variant that will produce useful results

- Replace an action in the system with an opposite action
- Make mobile parts immobile, or vice versa
- Turn an object or system inside-out or upside-down
- Substitute an internal action for an external one, or vice versa

Separate (1)
Divide and then recombine in a more efficient way

- Replace a one-piece system with a partitioned one
- Exclude an element by transferring its function to the remaining elements
- Exclude an auxiliary function by transferring it to the remaining ones
- Use a temporary or disposable object instead of an expensive permanent system
- Perform a function partially rather than completely (80/20 rule)
- Take out from the system the part that performs the useful function

Improve a useful function (part B)

Integrate (5)
Create synergy by consolidating functions

- Merge subsystems that perform similar functions
- Build a Bi- or Poly-System by integrating two or more systems into a new synergistic system
- Combine a function with an opposite one for improved control
- Repeat a function several times (19)
- Conduct sequential functions concurrently, simultaneously
- Process multiple objects together
- Use a mediator to transfer action or energy

Dynamism (15)
Make the system more dynamic, more flexible and adjustable

- Transform static relationships between elements into flexible, and changeable relationships
- Introduce a dynamic, modifiable element into the system
- Dynamic Poly-System: replace a rigid unchangeable system with a set of flexibly connected discreet elements
- Transform a statically stable system into a dynamically stable system

Control
Make the system more controllable

- Introduce feedback into the system
- Enable the system to self-adjust to changing operating conditions
- Replace an uncontrollable system element with an element that is controllable
- Use information obtained from changes in substance properties to increase system ideality
- Use flows from a system as an informational resource
- Use flows passing through the system as an informational resource

FIGURE 13.3 (CONTINUED)
Series of nested I-TRIZ operators.

Combine Ideas into Concepts and Prioritize for Implementation

Ideas from guided process are first organized and then combined into useful concepts (where two or more ideas implemented together provide a better innovation). These concepts are then prioritized for implementation. The ETI case study below provides an example.

Case Study: ASQ's Education and Training Initiative (Smith, 2008)

Abstract and Introduction

The outcome of an October 2005 meeting of member leaders of the ASQ determined that, "There is a shared belief that ASQ could create a much more responsive system to provide the quality community with education and training." A team was tasked to, "Start with a clean slate and develop a vision and long view for our training and education efforts. Develop something modern, something responsive to the needs of customers and members, and something that makes the best use of ASQ's tremendous talents and capabilities."

Four ASQ staff members were trained with both Ideation TRIZ Knowledge Wizard® and GOAL/QPC Seven Management and Planning Tool software. TRIZ was used to better understand the issues, to model the system and find leverage points for system change, and to develop solution ideas. Quality tools were used to establish and prioritize system vision elements, create a vision, identify and prioritize system inhibitors, and to prioritize the solution ideas.

The integration of TRIZ innovation and quality methods provided a very efficient and effective process for planning improvement actions.

Project Strategy and Preparation

The strategy for this project called for the creation of two levels of cross-functional teams. First, a steering team composed of leaders of various society activities who are stakeholders in the education and training system. Their roles would be to

- Promote the education and training vision
- Assist in staffing the working teams
- Contribute ideas on working team direction/initiatives
- Review working team progress; provide feedback

- Assist with resource needs; help remove roadblocks
- Achieve consensus with working team recommendations
- Champion implementation of action items

Secondly, working teams would be formed consisting of members and staff who have more detailed knowledge and time to perform the needed tasks.

It was decided to kick off the effort with a two-day meeting consisting of invited participants representing various functions of the education and training system. Twenty-three participants representing the perspectives of corporate customers, the ASQ board of directors, the ASQ Education and Training Board (ETB), ASQ management and staff, sections, divisions, instructors, training material suppliers/providers, and an education expert (with no previous knowledge of the ASQ) met in Milwaukee in February of 2006. The purpose of this meeting was to tap their knowledge to establish a draft vision for education and training, identify and prioritize inhibitors, formulate a system relationship map, and identify and prioritize solution options. The membership and tasks for the steering and working teams were outcomes of this initial effort.

The following actions were taken to prepare for this meeting:

- Four ASQ staff members were trained with software to facilitate subteams with the Seven Management and Planning Tools (7MP software from GOAL/QPC) and nontechnical TRIZ (Knowledge Wizard software from Ideation International Inc.). As an interesting and important side note, one trained staff member became proficient in TRIZ and has been called on to facilitate a number of various projects within the ASQ organization. In 2007 she was promoted into ASQ management.
- A core group met and prepared drafts of the Knowledge Wizard's ISQ and a map of system relationships using Knowledge Wizard's Problem Formulator®. These drafts were later circulated and enhanced by the full team.

Innovation Situation Questionnaire

Ideation's ISQ is a product of many years of TRIZ research and practice on the part of TRIZ masters Boris Zlotin and Alla Zusman and their colleagues and associates, who found that many teams failed because they incompletely understood the problem they were dealing with, and when they jumped to solutions they wound up implementing solutions that did not actually solve the problem they wanted to address.

So Zlotin and Zusman carefully crafted a series of questions to force teams to think about what is really at the heart of the problem they are trying to address. Teams find answering and documenting the questions in the ISQ to be time consuming and tedious. Indeed, thinking is hard work! It takes an average of six hours of serious thinking to draft an ISQ, and it is very tempting to take shortcuts. However, thinking, discussing, and providing answers to the ISQ questions frequently result in self-discovery. The author has witnessed many teams that, after an ISQ exercise, formulate a very different problem to solve than the one they started with.

At the beginning of the ETI ISQ, the emphasis was on revenue generation, with the goal of "Selection, development, and delivery of educational events that enhance the capability of members and member companies and generate revenue for sections, divisions, partner providers, and HQ." As the team got deeper into the ISQ, the emphasis shifted to, "The development and delivery of educational events involving quality technology, concepts, and tools to enable members to improve themselves and their world."

Completing the ISQ Provided the Following Benefits

- Open and honest sharing of information from different team members with different points of view: the ISQ provides an opportunity for all team members to contribute information and perspectives in a nonthreatening way. With no sides or points of view to defend, all the information and perspectives become available to the whole team.
- Enhanced communication as team members share stories, query, and learn from each other.
- Documentation and sharing of existing relevant information, data, and research (often as attachments to the ISQ).
- New team insights and alignment on what the team is really trying to solve.
- Insights into resources that are available in the supersystem, system, or subsystems that can contribute to the solution.

- The generation of new ideas due to dialogue and insights associated with answering the ISQ questions. Thirty-four ideas were generated and documented as a result of working on the ETI ISQ.
- Team formation and alignment: As the team works on the questions, they become acquainted with each other and more aligned in what they are trying to accomplish.
- Productivity: Prework on the ISQ enabled the team to move quickly and efficiently through the I-TRIZ process, so that they accomplished a great deal with the face-to-face time they spent together.

As an example and an encouragement to other teams, an abridged version of ETI's ISQ is shown in the "I-TRIZ Process ISQ for ASQ's ETI" portion of this chapter.

Ideal Vision

After reviewing and refreshing the ISQ documentation, the team began an exercise using two (affinity and relations diagrams) of the Seven Management and Planning tools to establish an ideal vision for ASQ's education and training effort.

The exercise began by asking the team members to imagine that the ETI effort is a great success, and that this group is meeting again a decade from now to celebrate this success. Each participant was asked to think of three things the team members are saying to each other, and to write each item on a separate Post-it note. The Post-it notes were then organized by grouping statements with similar affinity, and titles were selected for each of these groups (Figure 13.4).

Next, GOAL/QPC's 7MP software was used to map relationships between the affinity diagram's major groups by drawing cause–effect arrows from one grouping to another (Figure 13.5). The number of arrows coming into and out of each group was counted in two ways: not weighted and weighted for strength of relationship. Groups with the most arrows going into them are more results oriented and therefore more closely related to the team's vision. By picking key words from these groups, the team drafted several vision statements. These statements completed a shift in thinking associated with the purpose of education and training, shifting from independent revenue generation to a collaborative enabling of others to improve themselves, their organizations, and their world. These

Stakeholder health
- ASQ is financially healthy...
- High stakeholder satisfaction

Collaboration
- Section training delivery = financial health
- Section, divisions, and HQ work together
- Collaboration replaces competition

Quality = business system
- Quality tools integrated and embedded
- Quality culture is pervasive
- The quality system is the business system

Variety of topics
- Breadth and depth of courses
- Product for sections, divisions, universities...
- Variety of training topics and deliveries

Variety of modalities
- Readily available in an individualized manner
- Customers have access when/how they want it
- Known for effective training all mediums

Career value
- Training as a complete process versus an event
- Learning curriculum encourages certification
- Develop and deploy many services/products

Knowledge transfer
We have successfully transferred knowledge of the quality masters to the next generation

Increase BoK in innovation
ASQ known for its leadership in innovation and sustainability

Consistent, high quality
- Trainers are exceptional
- Provide uniform material, society-wide
- Education is fun, effective and interesting

Value perception
- Quality professionals become management
- Individuals are able to improve performance
- Employers can see the ROI

Marketing
- Easy (cheap) for local non-ASQ members
- ASQ is an approved government provider
- I can use ASQ's web site for learning options

Improved access
More individuals have access locally to more training for less money/time expended

Cost of delivery
- Affordable on-demand training courses
- Corporate economies of scale (reduced $)
- Get breakeven class size down to 3

#1 quality training source
- ASQ is *the* source for quality learning
- ASQ image in training is everything!
- ...a structured training curriculum that is *world class go* to education in quality

Loyalty
- Confidence due to ASQ *Seal of Approval*
- Students return and recommend
- Members brag about consistent quality

Globally available
- All ASQ training is available globally
- Worldwide understanding of quality BoK
- Huge enrollments across many languages

Understanding VOC
Needs assessment leads to defining and implementing solutions that are immediately sought because of their value

Cost free
Members have free access to training products
Sales to non-member units supports activity

Increased customer base
- Key offerings engage and grow members
- 2016 – there are 10 times more participants
- 80% of ASQ members are touched

FIGURE 13.4
Relations diagram of ideal vision elements.

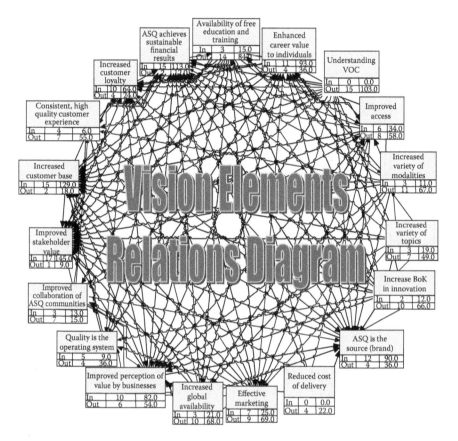

FIGURE 13.5
Vision element relations diagram.

drafts were later shared with the steering team, who used them to craft the following vision statement:

> ASQ's education system enables anyone, anywhere, anytime to improve themselves, their organizations and their world through quality concepts, technologies, and tools.
>
> ASQ's educational system is the seamless integration of HQ, member communities, education professionals, and members.

Inhibitors

The team was then tasked to identify inhibitors that would keep them from accomplishing this vision. An inhibitor affinity diagram was created. These inhibitor groups were prioritized using a relations diagram to find inhibitors that had more influence than others (prioritized using the number of arrows coming out of each group in the relations diagram).

In the ETI system, the top three inhibitors were lack of member unit (headquarters [HQ], sections, and divisions) collaboration, limited resources, and no master plan for the ETI system (see Figure 13.6). The team then studied and refreshed an earlier draft mapping of system relationships.

System Relationships

Ideation's Problem Formulator is a tool for mapping relationships in a system. Teams use this tool to understand and document cause–effect or *causal loop* interactions between useful (shown as square boxes) and harmful (shown as rounded boxes) elements in their system. The ETI system map is shown in Figure 13.7. For many team members, this was the first time that they thought about their issues in the context of a larger system. They were surprised by the scope and complexity of the system they were operating in, and could better appreciate why making progress is so difficult.

Ideation's Knowledge Wizard software uses system maps to identify leverage points for system change. The software seeks patterns similar to those shown in Figure 13.8, which shows a physical contradiction on the left and a related technical contradiction on the right. Functions are conceptual ideas linked with the actions or workings of a system. Contradictions represent functional conflicts, and there are two types. A function that conflicts with itself (e.g., Sally wants the room temperature to be 75°F, while Sam wants it to be 68°F at the same time) is called a physical contradiction. A function that conflicts with another function of a system (you want your automobile to have great acceleration and great fuel economy, but rapid acceleration reduces fuel economy) is called a technical contradiction.

One of Altshuller's great insights was that physical and technical contradictions are linked, with a physical contradiction underpinning a functional contradiction as shown in Figure 13.8. He called this insight OTSM, a Russian acronym for *theory of strong thinking*. For example, in the ASQ ETI system, there is a physical contradiction associated with HQ, sections, and divisions all conducting independent training. We want them to train in this way because they generate revenue. But we do not want them to train in this way because there is large variation in quality of event. In this

No master plan

- No overall master plan for training products
- Lack of understanding how the education and training system works (efficient or effective) society wide
- Training effort is a list of stuff and is not part of a roadmap or model
- The infrastructure is not in place to deliver training to the remote regions of the U.S. and the world where training is needed

Ineffective marketing and sales

- Not enough involvement from business community
- Poor marketing/inconsistent marketing and sales activities
- There is no unified system to promote training

Variation in program design

- Inconsistent quality of product from multiple deliveries (ASQ brand)
- Standards for course design are not systemic

Limited scope of offerings

- Lack of well developed distribution methods (face to face, e-learning, etc...)
- New technology is not yet mature or understood (e-learning, podcasts, etc.)
- Thinking of training as a *course* constrains better innovations in training

Not understanding customer needs

- We do not understand the needs of many types of customers—focus too narrow
- Not designed to meet customer needs
- Lack of focus on the customer
- Needs assessment is too often based on *wants* rather than true needs

Lack of HQ and member unit cooperation and coordination

- Lack of role delineation between HQ, sections, and divisions
- Lack of cooperation, coordination, and communication
- The training is torn between varying interests: sections, divisions, HQ, and consultants
- Lack of consistent communication between HQ/sections

Lack of perceived value

- Companies do not see a pay-off for high priced quality training
- Are the customers getting what they want? Are they applying what they learn? Are they getting results?

Limited resources

- Limitations of member leader time and effort
- Training offerings are all piecemeal, no cohesive framework to integrate or differentiate offerings

Intellectual property conflicts

- Ineffective supply chain (course provider) management
- Local developer of course leaves and takes course with them

Internal competition

- A structure that encourages competition among stakeholders
- Competition with and between sections, members, HQ, etc.
- The desire of everyone in the system to benefit economically

Inconsistent learning experience

- Quality of instructor
- One bad experience can deflect customer away from all offerings
- Quality of training is inconsistent

High price of programs

- Cost is said to limit participation
- They remain out of reach for many quality professionals in terms of cost/affordability

Lack of assessment of external competition

ASQ is competing with self-taught, non-experts, in-house facilitators that provide a look-alike alternative

FIGURE 13.6
Affinity diagram of ETI inhibitors.

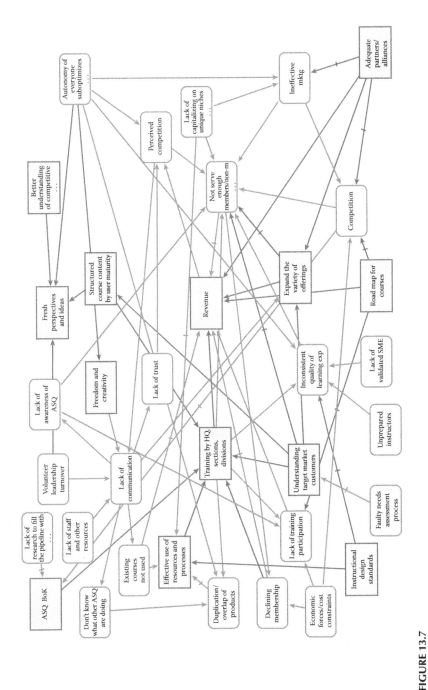

FIGURE 13.7

ETI system map showing relationships between physical and technical contradictions.

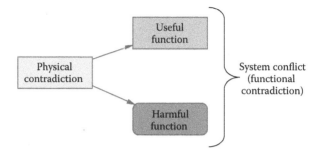

- Resolve contradictions/conflicts
- Make useful functions more useful
- Make harmful functions less harmful

FIGURE 13.8
Approaches for system improvement.

example, the technical contradiction is a current system conflict between revenue generation and variation in quality of event.

In Figure 13.8, there are three generic approaches for improving a system:

- Resolve the contradiction(s).
- Make useful function(s) more useful or find an alternative to this function free of the current conflict.
- Make harmful function(s) less harmful or eliminate the harmful function(s) altogether.

Ideation's Knowledge Wizard software provided the team an exhaustive list of approaches for system improvement using the team's formulated system map.

From the ETI system map, the software identified 47 possible approaches for improvement. The team reviewed these approaches and selected the following for further work: (Note: The following numbering system applies to the characteristic number that is specific to situation under study and as such is not sequenced in order.)

2. Try to resolve the following contradiction: The useful factor [the] (Training by HQ, Sections, Divisions) should be in place in order to provide or enhance [the] (Revenue), and should not exist in order to avoid [the] (Inconsistent Quality of learning experience), (Not serve enough members/nonmembers), (Duplication/overlap of products), and (Lack of training participation).

2.1. Try to apply Universal Operators to circumvent the contradiction.

5. Try to resolve the following contradiction: The useful factor [the] (Revenue) should be in place in order to fulfill useful purpose and should not exist in order to avoid [the] (Perceived Competition).

5.1. Try to apply Universal Operators to circumvent the contradiction.

6. Find a way to eliminate, reduce, or prevent [the] (Perceived Competition) in order to avoid [the] (Not serve enough members/ nonmembers) and (Lack of communication), under the conditions of [the] (Lack of trust), (Autonomy of everyone suboptimizes the whole (know or care)), and (Revenue).

6.1. Find a way to benefit from [the] (Perceived Competition).

6.2. Find a way to decrease the ability of [the] (Perceived Competition) to cause [the] (Not serve enough members/nonmembers) and (Lack of communication).

8. Find a way to eliminate, reduce, or prevent [the] (Lack of communication) in order to avoid [the] (Lack of trust), (Existing courses not used), (Don't know what other ASQ are doing), and (Lack of awareness of ASQ), under the conditions of [the] (Volunteer Leadership turnover), (Lack of Staff and other Resources), (Perceived Competition), and (Freedom and Creativity).

8.1. Find a way to benefit from [the] (Lack of communication).

8.2. Find a way to decrease the ability of [the] (Lack of communication) to cause [the] (Lack of trust), (Existing courses not used), (Don't know what other ASQ are doing), and (Lack of awareness of ASQ).

11. Find a way to eliminate, reduce, or prevent [the] (Inconsistent quality of learning experience) in order to avoid [the] (Not serve enough members/nonmembers), under the conditions of [the] (Unprepared instructors), (Lack of validated SME), (Autonomy of everyone suboptimizes the whole (know or care)), and (Training by HQ, Sections, Divisions).

11.1. Find a way to benefit from [the] (Inconsistent Quality of learning experience).

11.2. Find a way to decrease the ability of [the] (Inconsistent Quality of learning experience) to cause [the] (Not serve enough members/nonmembers).

16. Find an alternative way to obtain [the] (Understanding target market customers) that offers the following: provides or enhances [the] (Training by HQ, Sections, Divisions), (Expand the variety

of offerings) and (Structured course content by user maturity), eliminates, reduces, or prevents [the] (Not serve enough members/nonmembers), and is not influenced by [the] (Faulty needs assessment process).

16.1. Find a way to increase the effectiveness of [the] (Understanding target market customers).

16.2. Find additional benefits from [the] (Understanding target market customers).

16.3. Find a way to obtain [the] (Training by HQ, Sections, Divisions), (Expand the variety of offerings), and (Structured course content by user maturity) without the use of [the] (Understanding target market customers).

16.4. Try to increase the effectiveness of the action of [the] (Understanding target market customers) toward reducing the harmful nature of [the] (Not serve enough members/nonmembers).

16.5. Consider modifying or influencing [the] (Not serve enough members/nonmembers) to improve its being eliminated, reduced, or prevented by [the] (Understanding target market customers).

24. Consider replacing the entire system with an alternative one that will provide [the] (Fresh perspectives and ideas).

24.1. Consider transition to the next generation of the system that provides [the] (Fresh perspectives and ideas), but which will not have the existing problem.

24.2. Consider enhancing the current means by which the primary useful function is achieved, to the extent that the benefits will override the primary problem.

24.3. Consider giving up the primary useful function to avoid the primary problem.

37. Find an alternative way to obtain [the] (Road map for courses) that offers the following: provides or enhances [the] (Revenue); eliminates, reduces, or prevents [the] (Lack of training participation) and (Competition).

37.1. Find a way to increase the effectiveness of [the] (Road map for courses).

37.2. Find additional benefits from [the] (Road map for courses).

37.3. Find a way to obtain [the] (Revenue) without the use of [the] (Road map for courses).

37.4. Try to increase the effectiveness of the action of [the] (Road map for courses) toward reducing the harmful nature of [the] (Lack of training participation) and (Competition).

37.5. Consider modifying or influencing [the] (Lack of training participation) and (Competition) to improve its being eliminated, reduced, or prevented by [the] (Road map for courses).

44. Try to resolve the following contradiction: The harmful factor [the] (Autonomy of everyone suboptimizes the whole (know or care)) should not exist in order to avoid [the] (Perceived Competition), (Duplication/overlap of products), (Inconsistent Quality of learning experience), and (Ineffective marketing), and should be in place in order to provide or enhance [the] (Freedom and Creativity) and (Fresh perspectives and ideas).

44.1. Find a way to increase the effectiveness of [the] (adequate partners/alliances).

44.2. Find additional benefits from [the] (adequate partners/alliances).

44.3. Find a way to obtain [the] (Revenue) and (Expand the variety of offerings) without the use of [the] (adequate partners/alliances).

44.4. Try to increase the effectiveness of the action of [the] (adequate partners/alliances) toward reducing the harmful nature of [the] (Competition) and (Ineffective marketing).

44.5. Consider modifying or influencing [the] (Competition) and (Ineffective marketing) to improve its being eliminated, reduced, or prevented by [the] (adequate partners/alliances).

47. Find an alternative way to obtain [the] (adequate partners/alliances) that offers the following: provides or enhances [the] (Revenue) and (Expand the variety of offerings), eliminates, reduces, or prevents [the] (Competition) and (Ineffective marketing).

Idea Generation Using System of Operators

The above work was the combined output of the prework and the first meeting day. The second day began by splitting the team into three groups. Three of the ASQ staff trained facilitators each took a third of the group, and were assigned to work a portion of the above directions with Knowledge Wizard's TRIZ system of operators.

The Knowledge Wizard's system of operators is another innovation from Boris Zlotin and Alla Zusman. On the basis of their many years of experience and a desire to provide more rapid and relevant access to applicable TRIZ research, principles, standard solutions, and laws/lines of evolution, they have restructured the classical TRIZ knowledge base into various nested series of operators organized around the laws by which systems evolve. An example of this is shown in Figure 13.9 (taken from the later version of Knowledge Wizard, version 3.2.0). A subteam working on approach 2 above would be guided by the software to begin systematic brainstorming using separation principles. By exploring these principles, they would generate and record ideas, and would begin moving deeper into how these principles could be accomplished. The software brings up examples for them to consider and stimulates their thinking to generate new ideas. The ETI subteams were given three hours to come up with solution ideas. A little more than 200 ideas in total were generated by these teams. (In general, team members were impressed with the number of ideas they generated. One team member remarked that if they were working on a topic like this in their company, they would be typically come up with 20 ideas!)

The solution ideas were organized in an affinity diagram, shown in Figure 13.10.

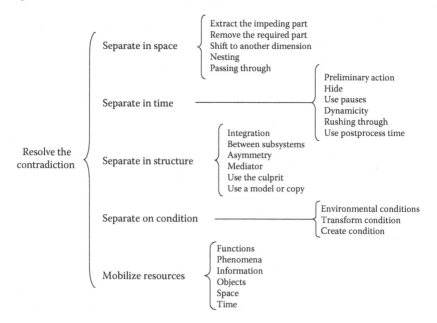

FIGURE 13.9

Example of Boris Zlotin's and Alla Zusman's system of operators.

Solution ideas

VOC research
- Create a mechanism to push good ideas, request for help
- Get on the phone or go to sections and ask what the customers need and want

Optimize electronic learning
- Move to more virtual/electronic communication
- Broadcast training opportunities from larger areas to the less served areas
- Package the e-learning to optimize learning

Optimize modular material design
- Create reuseable modules
- Develop huge library of modules and examples developed for different target customers
- Materials online so we would not have to print materials

Create a for-profit subsidiary
- Create a for-profit subsidiary for training and education

Optimize product development
- Develop a learning management system
- Develop materials in a modular way

Paid regional field service representatives
Professional staff around the country to sustain consistency of products and understand customer needs

Research
- Corporate research collaboratives
- Do more thorough research with identified target markets

Revenue
- Do not have independent discrete revenue generators
- Revenue issues need to be dealt with—all of the parties need to participate

Understand business trends and goals
- Develop processes where materials are translated and updated
- Inputs from the organizations that employ quality professionals
- Work with the organizational members to identify needs and issues to understand how ASQ can help to achieve results

Some offerings for free
If program offerings for free it may offer value to the membership—market training as a membership benefit

Learning event communications
- Design the process so that communication occurs, and all the right people are involved
- Develop a master schedule of speakers, trainers, courses, conferences

IP policy development
- Determine and define ownership
- How can ASQ acquire the IP and have at least co-ownership rights?

Leverage certification programs
- Create a freshman, sophomore, junior, senior sets of training that leads to certification
- Make the recertification requirements more rigorous (involving ASQ training or education)

Leverage stories from customer organizations
- Develop corporate entities as the recipients and then disseminators of the training
- Involve, partner with companies and use their success stories in conferences and courses

Measurement tool for individual/corporate quality maturity
- Develop a diagnostic tool that assesses quality growth
- Ask participants to beta-test diagnostic tool

Mega-events, piggy back off similar logistics coordination
Integrate the division conferences into the world conference

Sales/marketing strategies
- Enhance the perception of quality, quality ideas, and quality tools
- More bundled courses for certificates

Utilize conferences/methods for material awareness/marketing
- Instructors also act as agents to sell ASQ offerings
- The conference becomes a forum where you could learn how to use the materials

ASQ-certified instructors and courses
- Establish standards for training and certify trainers by course
- Do not put ASQ brand on courses that aren't certifies by ASQ

ASQ develops and licenses materials, others deliver
- ASQ as an organization will franchise out its learning materials
- ASQ could provide one set of training materials of all to use

Benchmark peers to improve process
- APQC-benchmark clearing house—diagnostic tool-share branding
- Benchmark other organizations to evaluate their structures. Learn from others

Optimize MU–HQ collaboration
- All registration comes through headquarters
- Encourage sections and division to collaborate with HQ training

Create partnerships to expand base and ease burden of logistics, marketing
- ASQ should think more like a university or establish relationships with colleges/universities
- Write articles for other publications to build intrigue

Develop and market integrated systems/solutions for corporations
- Do not deal with individuals, deal with the organizations that employ them
- Look for customers in non-quality roles and professionals within the target markets

Develop materials that do not need SMES
- Develop the cadre of local facilitators that do not need to be a SME
- Trained facilitators that can lead a discussion

Win—win incentives for all participants
- Create incentives for SMEs to give co-ownership for the IP
- Incentivize sections and divisions to put on certified training

FIGURE 13.10
Affinity diagram of ETI solution ideas.

Solution Prioritization

The Seven Management and Planning tools were used to prioritize the ideas using four different points of view:

- A matrix indicating the solution's impact (none, small, medium, or large) on the top three vision elements and inhibitors
- The above matrix, with vision elements and inhibitors weighted with values from the relations diagram
- A matrix data analysis assessment of solution importance versus how well solution elements are already being implemented in the organization (This approach prioritized solutions that are above average in importance and below average in implementation, and is shown in the bottom portion of Table 13.1.)
- A relations diagram of the solutions themselves, prioritizing solutions that have the greatest impact on other solution ideas

The matrix of solution ideas versus vision and inhibitor elements (both weighted and not weighted (see Table 13.1) was completed by the working team at the end of the second day, which was the last day of two-day meeting. The author with the facilitators of the subteams completed the second two methods for prioritization and summary results in a matrix, shown in Figure 13.11. This work was shared with the steering team in an all-day meeting. After considerable discussion, the steering team selected the following solution ideas for implementation (see Figure 13.11):

- Voice of the customer (VOC): To better understand and institutionalize processes and methods at HQ to better understand customer's needs, wants, and delights.
- Management structure: To benchmark other organizations, study the current competitive organizational structure, and establish design principles for a new, more collaborative organizational system.
- Modular development: To develop a modularized library of course materials in collaboration with sections and divisions so that the modules can be quickly mixed and matched to quickly create customized curriculums that meet consumer needs.

TABLE 13.1

Solution Effect on Key Vision/Inhibitor Elements (Weighted/Not Weighted)

	Weight	Optimize Modular Material Design	VOC Research	Paid Regional Field Service Representatives	Optimize Product Development	Optimize Electronic Delivery	Improve Communications Around Learning Events	Win–Win Incentives	Optimize Member Unit/HQ collaboration	Leverage Certification Program	Develop and Market Integrated System for Corporations	Leverage Stories from Customer Organizations	Sales Marketing Strategies	ASQ Certifies Instructors and Courses	Create a Subsidiary	Measurement Tool for Individual/Corporate Quality Maturity	Develop Materials that Do Not Need SMEs	HQ Develops and Licenses Materials	Mega Events	Create Partnerships to Expand Base and Base Logistics Burden	Benchmark Peers to Improve the Process	Understand Business Trends and Goals	Develop IP Policy	Some Offerings for Free	Totals
Weighted totals		2451	3483	2709	2451	2451	2451	2709	1581	2709	3483	2709	1805	1581	1805	1839	548	774	903	1419	903	1161	0	1709	43634
Totals		2341	2349	2343	2341	2341	2341	2343	1465	2343	2349	2343	1431	1465	1431	1467	552	780	781	269	781	783	0	561	35200
Stakeholder value	113	9	9	9	9	9	9	9	9	9	9	9	1	9	1	9	1	3	3	1	3	3	0	1	134
(weighted)		1017	1017	1017	1017	1017	1017	1017	1017	1017	1017	1017	113	1017	113	1017	113	339	339	113	339	339	0	113	15142
Sustainable financial results	145	9	9	9	9	9	9	9	3	9	9	9	9	3	9	3	3	3	3	1	3	3	0	3	136
(weighted)		1305	1305	1305	1305	1305	1305	1305	435	1305	1305	1305	1305	435	1305	435	435	435	435	145	435	435	0	435	19720
Increase customer base	129	1	9	3	1	1	1	3	1	3	9	3	3	1	3	3	0	0	1	9	1	3	0	9	68
(weighted)		129	1161	387	129	129	129	387	129	387	1161	387	387	129	387	387	0	0	129	1161	129	387	0	1161	8772
Lack of HQ–MU cooperation	60	9	3	3	3	1	9	3	9	3	0	0	3	3	3	0	1	3	3	3	1	3	9	1	76
(weighted)		540	180	180	180	60	540	180	540	180	0	0	180	180	180	0	60	180	180	180	60	180	540	60	4560
Limited resources	66	9	9	9	9	9	1	3	9	1	1	1	9	3	3	3	9	3	3	9	3	1	3	1	111
(weighted)		594	594	594	594	594	66	198	594	66	66	66	594	198	198	198	594	198	198	594	198	66	198	66	7326

(Continued)

TABLE 13.1 (CONTINUED)

Solution Effect on Key Vision/Inhibitor Elements (Weighted/Not Weighted)

																							31
No master plan	83	1	1	3	1	1	1	1	1	1	1	1	1	3	1	3	1	1	1	1	3	1
		83	83	83	83	83	83	83	83	83	83	83	83	83	83	83	83	83	83	83	83	83
Totals		1153	787	789	787	665	617	385	1153	251	68	787	385	202	665	387	385	787	263	251	753	129
																						2573
Weighted totals		1217	857	1023	857	737	689	461	1217	329	149	857	461	281	737	627	461	857	341	329	987	209
																						12104
Overall total	3494	3136	3132	3128	3006	2958	2728	2618	2594	2411	2218	1850	1669	1217	1167	1166	1056	1044	1034	753	690	14459
Overall total (weighted)	3668	4340	3732	3308	3188	3140	3170	2798	3038	2858	2662	2042	2120	1285	1401	1364	2276	1244	1490	987	1918	47304
Overall percentage	7%	7%	7%	7%	6%	6%	6%	6%	5%	5%	5%	4%	4%	3%	2%	2%	2%	2%	2%	2%	1%	58093
Importance	4	5	4	5	4	3	5	5	4	4	5	4	2	2	5	2	5	5	3	4	2	3.91304
Implementation	1	2	2	3	3	2	2	2	3	3	3	1	1	1	3	1	3	3	3	2	3	2.17391
Ease of implementation																							
Resources required																							
Greater than average importance	x	x	x	x	x		x	x	x	x	x	x			x		x	x		x		
Greater than average implementation	x	x	x		x	x			x	x	x		x		x	x		x		x		

Solution Prioritization

by:

- Importance versus
  implementation

- Effect on other
  solutions

- Effect on vision and
  inhibitor elements
  (weighted and not
  weighted)

	Importance versus level of implementation	Effect on other solutions	Effect on vision elements/inhibitors (not weighted)	Effect on vision elements and inhibitors (weighted)
Leveraging certification			X	
IP policy	X			
Subsidiary		X		
Certify instructors–courses	X	X		
Measurement tools of maturity	X			
Product development		X	X	X
HQ–MU collaboration	X			
Win–win incentives	X			
Integrated systems	X			X
Modular material design	X		X	X
Sales and marketing–business trends			X	X
Regional service reps	X	X	X	X
VOC research	X	X	X	X
Benchmarking		X		
Mega events		X		

FIGURE 13.11
Prioritization of ETI solutions from four points of view.

Progress

In 2006, volunteer team leaders and team members met on a regular basis
to work on the above three solution ideas. They completed phase 1 of the
work and shared their results with all member leaders at the 2007 ASQ
World Conference in Orlando. Phase 2 expanded the work of these three
initial teams into seven teams that were tasked with

- Continuing the work of institutionalizing the VOC into sections and
  divisions
- Creating an organization-wide ETI process owner, providing VOC
  and ETI information packets to each section and division in time for
  their business planning process, and improving communications by
  establishing communities of practice involving education chair and
  VOC representatives in each section and division
- Curriculum roadmaps: Recommended quality curriculums for individuals and organizations by market segment and job function

- Learning event best practices and recognition: Checklists and guidelines for actions to be taken before, during, and after a learning event, reducing variation in quality of event
- Intellectual property: Creating standardized contracts with providers that support modular development of existing materials
- Expand the role of the ETB: To define a vision of how the ETB of the future will operate, including roles of peer course review, VOC, curriculum strategy and development, measures of success, and instructor validation
- Modular course creation and delivery: To pilot a new HQ process in order to develop modularized courses in collaboration with sections and divisions with appropriate revenue sharing

The work of these seven teams resulted in the implementation of a new organization structure and processes.

Conclusion and Summary Remarks

ASQ's education and training system is exceedingly complex and difficult to manage and change. The combination of quality and TRIZ innovation methods produced a very effective and efficient means of finding the right strategy for organizational transformation and improvement. TRIZ was used to better understand the issues, to model the system and find leverage points for system change, and to develop solution ideas. Quality tools were used to establish and prioritize system vision elements, create a vision, identify and prioritize system inhibitors, and to prioritize the solution ideas.

It is interesting to note that although three of the TRIZ solutions were selected for initial work, as slow and steady progress continued, more and more solution ideas were pulled in for development and implementation.

It is also important to note that the same inhibitors that affect ASQ's education and training system also operate in every other aspect of the society. ETI is building a road through these inhibitors. This road, once built, can streamline all other aspects of ASQ business.

I-TRIZ Process ISQ for ASQ's ETI

1. Brief Description of the Problem

- Feedback from members through stakeholder dialogues and needs research states loudly that members want *training and education* and either want it locally or electronically, and at low cost.

- Customer research indicates *education and training* as a strong determiner of member value.
- Sections wish to train, and desire access to a variety of training materials from HQ or from other sections.
- Redundant effort—similar training material may be developed by ASQ HQ and different sections and divisions.
- Reaching ASQ's vision (making quality a global priority, organization imperative and personal ethic, etc.) is enabled by equipping people the knowledge and skills to practice quality.
- ASQ utilizes a limited array of learning offerings—a broader array could engage a greater spectrum of learners. (For instance, ASQ learning assumption is that there must be an instructor—which is not true.)
- Perception that sections, divisions, HQ, and member companies all compete with each other for training dollars limits data sharing, communication, and collaboration.
- Insufficient member engagement with training, especially in the target markets: service, education, health care, and international.
- Biggest membership is in manufacturing, a section of the US economy that is rapidly disappearing.
- Companies do not allow people to travel like they used to—spurs interest in e-learning and local opportunities.
- Perceived *sweetheart* deals with individuals; issuing of *no-bid* contracts for training services.
- Available training materials developed for section use are not being used.
- Section leadership turns over so frequently that it is difficult to keep everyone informed and maintain constancy of purpose.
- Wide range of quality of training, especially with volunteers—customer sees ASQ, not divisions, sections, HQ, etc.
- So many variables; each section and division operates differently and may change each year as the leadership changes.
- Sharing knowledge while protecting ASQ intellectual property.
- Enabling members outside the United States to use and benefit from ASQ's learning offerings.
- There is a growing interest of companies to have *in-house training*.
- Research indicates that the cost (high cost) is a concern of members in obtaining access to learning offerings.

Need to better understand and manage the system elements such that the performance of the system is optimized as a whole to the benefit of members, sections, divisions, member companies, and HQ. Need to incentivize a system that fosters collaboration and cooperation, with each element sharing a portion of a larger pie. Need win/win/win for everybody…

What are the roles of each element? What is the best business model for an association? What are the sources of products and innovation? Who is responsible for strategic thinking and direction? To what extent is training linked to individual performance? Is there a strategic link from individual performance to outcomes like customer satisfaction?

2. Information about the System

2.1 System Name

Selection, development, and delivery of educational events that enhance the capability of members and member companies and generate revenue for sections, divisions, partner providers, and HQ.

- There are two systems within ASQ. The education material, courses, and conferences developed by or in conjunction with the ASQ HQ and those developed and delivered by sections, divisions, and individual ASQ members. In addition, there is a plethora of consulting companies and individuals who also provide these services.
- *History.* Initially the training and conferences were developed by the larger sections, for example, Detroit, Chicago, and Boston. As ASQ and the body of knowledge grew, divisions developed and ASQ(C) took more of a role in the education process. (Comment in 3.3 bullet #3.)

2.2 System Structure

The board of directors oversee the activities of the society including the education and training board, the annual conference board, and section/division boards.

The ETB oversees public courses, e-learning, and self-directed learning for HQ. There is no comparable function for section and division courses.

The annual conference board oversees the World Conference. Divisions provide their own oversight.

HQ, sections, and divisions are independent and may compete for training dollars.

The competition issue may be a perceived issue versus fact. Also, ASQ is pursuing organizational training in addition to the public offerings.

Our information system has very limited data on the activities, engagement, and satisfaction of member unit-offered learning events.

The system is very structured at HQ with detailed processes but assumed to be very ad hoc at section and division levels. The annual churn of member unit leaders makes consistency a challenge. Consumers see all as ASQ, with the system only as strong as the weakest link.

Suppliers: instructors, program developers, and training providers (companies). New suppliers come in by responding to a call for proposals or by suggesting a new idea and having a new idea accepted by the ETB.

2.3 Functioning of the System

The development and delivery of educational events involving quality technology, concepts, and tools to enable members to improve themselves and their world.

- The ETB approves all course titles, and ensures content validity through peer reviewers who are provided by the divisions.
- HQ determines cost, venue, and mechanism for the event (everything operational).
- The annual conference board oversees the development of programs for the ASQ World Conference. All papers are reviewed by division subject matter experts except for papers submitted by the divisions themselves.
- Sections develop or acquire and deliver their own local training/conferences, as needed, often partnering with trainers in their section or with local higher-education institutions.
- Divisions develop or acquire and deliver their own industry training/conferences as needed.
- Pricing is all ad hoc, with each provider determining the price.
- There is obvious gross inefficiency in a system that has as many as three units of the same organization developing and delivering learning offerings and no sharing of results about success or failure, what is working and what is not.

HQ training receives feedback from the participants at the end of an event, and working toward three months following the event. Each section/division may have their own feedback mechanisms with data kept internally.

2.4 System Environment

Supersystem: customer care, IT, distribution center, event management, financial management, marketing services, market management, production services, community care, HQ partners, sales team, website, and publications.

Competition: colleges, universities, member companies, other professional societies, training providers, member units, and companies providing their own training.

3. Information about the Problem Situation

3.1 Problem that Should Be Resolved

Make it easier for HQ, sections, and divisions to fulfill their training missions while reaching more people/making more money with less effort. Optimize the training and education system.

(Executive director comment: "I'm not sure making more money needs to be so prominent a feature. It is a natural outcome of succeeding—the driver needs (in my opinion) meeting member needs and increasing the capability of the quality community.")

A failure or drawback must be corrected:

- Perceived competition between HQ, sections, divisions, and their alliances.
- Lack of trust (sweetheart deals, perceived conflicts of interest, us versus them, confidence HQ can deliver the training, no bid contracts, etc.).
- Inconsistent quality of learning experience between and within HQ, sections, and divisions.
- Not serving enough of the membership population.
- Not reaching and serving enough of the nonmember community.
- Target markets: what do service, health care, education, manufacturing, international need?
- HQ developing courses for sections/divisions that are not being used.
- Instructors who come to teach unprepared.
- Expand the variety of learning offerings and modes made available (venues/experiences).

- Make learning offerings available to non-North American members.
- Fix the inefficiency and duplication of development initiatives.
- Organizations that advertise, exhibit, sponsor, and pay for access to members in order to market their products and services see ASQ (a member organization) competing with them—redefine the meaning of alliances and competitors.
- Defined course content—targeted levels of maturity (road map).
- Road map of what courses to take to what?
- Published instructional design standards to achieve consistency of standards across all member units.
- What can HQ do for sections that are affordable and needed?
- Cost to the participant.
- What does ASQ do that is unique?
- Not enough competitive assessments by product line—including market share.
- Meaningful communication between HQ, sections, divisions, and corporate accounts.
- Lack of awareness of ASQ as a provider.
- Tools not properly named/advertised to the best marketing advantage.
- Faulty needs assessment process.
- Members do not feel valued by HQ.
- ASQ does not leverage ASQ higher-level members as part of its sales force.
- Autonomy of everyone suboptimizes the whole (know or care).
- System-wide lack of clear understanding of mission and scope.
- Not disciples of quality.
- Lack of incentive structure in individual member organization to recognize the value of training.
- ASQ is not a known benchmark for any of the quality practices.
- Lack of capitalizing on unique niches (Baldrige, certification, etc.).
- Lack of research to fill the pipeline with new offerings (academic research).
- Policies may support competition as opposed to rewarding cooperation.
- Cannot provide course content in the time frame needed.
- Fact-based decision making.
- Inadequate identification and maintenance of training partners/ alliances.

An operation must be improved:

- Market response, especially to target markets.
- Turning market data into useful products.
- Utilize resources in system to make it easier to operationalize and deliver training/education.
- Grow revenue potential for each stakeholder (HQ, sections, divisions, instructors)—lots of win/win/win scenarios.
- Communication and collaboration with sections, divisions, instructors, and corporate accounts when the local leadership may turn over frequently.
- Expectations of sections and divisions, their responsibility for leadership transition and communication among themselves.
- High levels of satisfaction for content quality, and instruction quality need to be assured by all stakeholders.
- System for collecting system-wide performance data.
- Gathering of personal data on individuals, as well as data on non-member companies and individuals.
- Better utilization of the environmental scanning process.
- Better trained leadership at the section and division levels.
- Improved leveraging of the niche areas (Baldrige, certification, etc.).
- More robust needs identification.
- Improved advertising/marketing of training.
- Coordinate with sections and divisions for revenue goals and market share.
- Inexpensive facilities at the section level to hold training events.

Information about an object's condition must be detected:

- Understanding the needs of target markets.
- Understanding needs of sections and divisions, both large and small.
- Do not know what sections and divisions are doing.
- The issues of revenue are different for sections, divisions, and ASQ. The sections need the revenue to support some of their activities and offer the members a reasonable cost suite of opportunities. What revenue is needed? What will the money be used for?

3.2 Mechanism Causing the Problem

- We are scattered around the world.
- We do not talk to each other or share any information.

- Scarcity mind-set with regard to training dollars.
- Compete with each other.
- Lack of shared objectives.
- Rapid turnover of local leadership.
- Assumption that *anyone can develop* learning events effectively and without cost.
- Assumption that success is economically driven (sections, divisions, and HQ all value money).

Where does the mistrust come from?

- Not share information.
- Start some rumors or unverifiable stories.
- Do not follow through on promises.
- Assume some things that are not true about money or where the resources are going.
- Suspicion that HQ is taking our money.
- Upset about sharing of membership dues carrying over to training and development.
- ASQ HQ not having quality professionals on staff contributes to the notion that *they* are not one of us.
- Don't feel they belong to all of ASQ; just a part of it.
- Perception of some that the board of directors does not represent sections and divisions the way they want to be represented—board members are in it for their own personal agenda.
- Each section and division totally independent and do not communicate—no accountability.
- Lack of a shared vision, and a lack of understanding or roles and responsibilities of how each part contributes to the whole.
- Stress associated with membership decline is being felt in every part of ASQ, puts more pressure to generate revenue to continue to live the way they have in the past—pressure translates to getting more revenue from training dollars.
- In addition, training does not generate as much revenue as it did in the past as companies cut back on training dollars—tendency to blame competition rather than climate for lack of training success.
- The board does task HQ with growing revenue, which creates purpose/goals/accountabilities for staff.
- In addition, volunteers who create the training have less time—working longer hours in the workplace as companies cut back on staff.

What is the needed balance within ASQ in terms of centralization/decentralization regarding training expertise, course development R&D and actual development of courses, the desire to achieve economies of scale, etc.?

3.3 Undesired Consequences of Unresolved Problem

- Grossly unmet need of customers for both access and satisfaction
- Continued competition, mistrust
- Decline in membership more likely to continue
- Decline in health of quality profession
- More difficult to meet financial goals
- Tarnished image
- Gaps in our knowledge of our actual reach and opportunity

3.4 History of the Problem

- Membership peaked in 1995–1996 and has been declining since.
- Follows trends of nation in the manufacturing sector.
- Training was not originally planned—it evolved to be everyone's opportunity.
- Bureaucratic inertia to respond to changes in the marketplace/economic factors.
- HQ developed certification material for all the sections back in the early 1990s to help people in the sections to become certified, but the sections found them more expensive and no more effective than their current materials supplier.

Last year 2005, HQ developed four courses for section use, based on section input—10 this year 2006, but sections are not using the materials/courses. Why?

- Communication problem?
- Maybe using but we do not know?
- How align with desire for *conference in a box* or *meeting in a box* or *course in a box*?
- HQ developed four webinars that were used by some of the sections. Two problems: communication and *firewalls* preventing the receipt of the online materials. Now looking at providing the material on CD or other. Security of intellectual property is an issue.

Do HQ courses need to be a break-even proposition?

September 11—Strong blow to training and conference attendance in an already declining market. Lead time for people signing up for events continued to decrease and every year is less.

3.5 Other Systems in which a Similar Problem Exists

Reward system for sections and divisions—divisions have to put on a successful conference to get award—perceived overlap/competition of conference dollars with HQ.

Can we learn anything from what England went through when they lost their industrial base at the turn of the century?

Are other organizations/associations seeing declining attendance at conferences/training?

3.6 Other Problems To Be Solved

The Institute of Electrical and Electronics Engineers (IEEE) does all their training electronically. Study and learn from their experience?

> To what extent are the terms training, learning, and performance differentiated? I assume that performance is a goal. Too often we consider training to be an *event* and learning to be a process, yet both are processes. The widespread use of e-learning changes our view of what we thought *training* was. To achieve the performance goal, we need to be more flexible with regard to the process to get us there. Thus, the long-standing *course-based* approach to training or learning is changing, especially outside the school-based environment. The bottom line—we need to be more innovative with regard to achieving successful performance.

4. Ideal Vision of Solution

HQ, sections, and divisions work seamlessly to anticipate, develop, and deliver to anyone, anywhere, implementable knowledge of quality technology, concepts, and tools with relevant examples to improve themselves and their world, whenever and however they need it.

The supplier of choice, ASQ offers readily available learning experiences to an ever-increasing number of loyal customers enhancing careers, organizations, and the world.

By creating a system that satisfies customers and partners, ASQ becomes the #1 source for quality training to an ever-increasing number of members and customers.

5. *Available Resources*

Financial:

- Current budgets of HQ, sections, and divisions
- Nonprofit development grants
- Special board funding
- Special support from member companies around areas of mutual concern
- Sponsorship of companies that want visibility
- Dues (could be thought of as a source for education development— members might pay more if access to education was included in dues)
- Foundations
- Other revenue streams such as books or software

Human:

- Testimonials and leadership of high-profile people
- Subject matter experts in sections and divisions
- Staff
- Liaison with other nonprofit activities
- Members and the talent/expertise each brings
- Partners
- Member leaders
- Volunteers
- Variety of reward and recognition systems, performance management systems
- ASQ China
- Lists of trainers and speakers

Technical resources

- Body of knowledge that goes back to the beginning of the quality movement
- Magazines, books, periodicals, conferences, existing courses

- Internet
- Technical staff: information technology systems, web systems, learning offering development, marketing
- Support system of registration accounting
- Rich member body knowledge and experience
- Local sections and divisions in place as delivery and development vehicles
- Variety of communication tools/medium
- Member/customer database
- Certification/accreditation systems

Other business assets:

- Unleased space in ASQ HQ building
- Forty-eight-hour turn-around customer inquiry, registration, delivery, systems, and processes
- Solid reputation
- Peer-reviewed materials—rich depth and integrity of books, articles, etc.
- Breadth of offerings: journals, magazines, books, courses, conferences, certification
- Market research and history of data

6. Allowable Changes to the System

The sky's the limit.
Drastic changes may be considered.
A whole system model is desirable.

7. Criteria for Selecting Solution Concepts

Solution concepts will be prioritized against vision elements and system inhibitors.

8. Company Business Environment

- See ASQ organization profile.
- Declining membership, competition (internally and externally).
- National transition from manufacturing to service economy.

- Amazing national debt and trade imbalance.
- Quality field is getting less attention in the last decade.
- People have less free time than they have in the past.
- Work/life imbalance.
- Aging/retiring of traditional quality professional.
- Many quality gurus have died—no ready replacements available.
- Manufacturing corporations have fewer resources to spend on education and training.
- People in corporations have less available time.
- Increasing resources for electronic communication.
- Keen/increased global competition fueling a need for innovation.
- Increased need/opportunity for standardization.
- Leveling the world's standard of living around the poorest-wage countries—elimination of middle class.
- Expanding breadth of topics in the quality body of knowledge.
- Growing global interest.
- Increased risk of terrorism and environmental catastrophe.

9. Project Data

References:

- Membership value leadership summit report—training ideas
- Membership value leadership summit action matrix
- Division affairs council/section affairs council meetings, Saturday, November 12, 2005
- Follow-up and cafe results

See the samples included in the section of this chapter entitled "How to Use the Tool."

SOFTWARE

I-TRIZ is supported by a TRIZSoft® software suite, as follows:

Main set:
- Innovation WorkBench®: A comprehensive professional tool for inventive problem solving.

- Knowledge Wizard®: A professional tool for inventive problem solving in nontechnical areas (business, management, marketing, logistics, etc.).
- Failure Analysis: A professional tool for revealing root causes of undesired effects (accidents, failures, production defects, etc.) and their elimination (prevention).
- Failure Prediction: A professional tool for predicting possible undesired effects and events (accidents, failures, production defects, etc.) and their prevention.
- Directed Evolution®: Provides a streamlined, accelerated, and controlled evolution of various evolving systems, including any kind of products, processes, services, technologies, etc. Businesses or organizations, countries, and human society as a whole can benefit.
- Invention Evaluation: A professional tool for objective evaluation of inventions.
- Invention Enhancement and Design Around: A professional tool for strengthening and increasing value of inventions/patents.

Optional Software

- Disclosure Preparation: A professional tool helping create an adequate invention description for further submission.
- Classical TRIZ Plus: A tool to aid in the application of Altshuller's original TRIZ methods.
- Ideation Brainstorming: A tool for solving problems in business, management, marketing, logistics, etc., in an individual or team setting.
- Problem Formulator: A tool for modeling system functions and finding directions for innovation work.

REFERENCES

Harrington, J., Voehl, F., Zlotin, B., and Zusman, A. The Directed Evolution methodology: A collection of tools, software and methods for creating systemic change. *The TQM Journal*, vol. 24, no. 3H, pp. 204–217, 2012.

Ideation International Inc. *IWB Self-Sufficiency in Inventive Problem Solving*, 3-Day Workshop Handbook, 2005.

Terninko, J., Zusman, A., and Zlotin, B. *Systematic Innovation: An Introduction to TRIZ (Theory of Inventing Problem Solving)*, 1998.

Smith, L.R. Case studies in TRIZ: ASQ's Education and Training Initiative. In: *TRIZcon2008, Proceedings of the Tenth Annual Conference of the Altshuller Institute for TRIZ Studies,* Kent State University, April 13–15, 2008.

SUGGESTED ADDITIONAL READING

Harrington, H.J., Fulbright, R., and Zusman, A. HU goes there. *Quality Progress,* September 2011.

Visnepolschi, S. *How to Deal with Failures (The Smart Way).* Ideation International Inc., 2008.

Zlotin, B. and Zusman, A. *Directed Evolution: Philosophy, Theory and Practice.* Ideation International Inc., 2001.

Zlotin, B. and Zusman, A. Creativity Tools, in *Encyclopedia of Statistics in Quality and Reliability.* In: F. Ruggeri, R. Kennett, and F.W. Faltin, eds. Chichester, UK: John Wiley & Sons Ltd., pp. 451–456, 2007.

Zlotin, B. and Zusman, A. Directed Evolution +B35. In: F. Ruggeri, R. Kennett, and F.W. Faltin, eds. *Encyclopedia of Statistics in Quality and Reliability.* Chichester, UK: John Wiley & Sons Ltd., pp. 544–547, 2007.

Zlotin, B., Zusman, A., and Fulbright, R. Knowledge based tools for software supported innovation and problem solving. Presented at TRIZCON, 2011.

14

Imaginary Brainstorming

H. James Harrington

CONTENTS

Together we are better than any one of us alone.

H. James Harrington

DEFINITION

Brainstorming is a technique used by a group to quickly generate large lists of ideas, problems, or issues. The emphasis is on quantity of ideas, not quality.

Imaginary brainstorming expands the brainstorming concept past the small group problem-solving tool to an electronic system that presents the problem/opportunity to anyone that is approved to participate in the electronic system. Creative ideas are collected, and a smaller group is used to analyze and identify innovative, imaginative concepts.

USER

This tool can be used by individuals, but its best use is with a group of four to eight people. Cross-functional teams usually yield the best results from this activity.

OFTEN USED IN THE FOLLOWING PHASES OF THE INNOVATIVE PROCESS

The following are the seven phases of the innovative cycle. An X after the phase name indicates that the tool/methodology is used during that specific phase.

- Creation phase X
- Value proposition phase X
- Resourcing phase X
- Documentation phase X
- Production phase X
- Sales/delivery phase X
- Performance analysis phase X

TOOL ACTIVITY BY PHASE

- Creation phase—Once an unfulfilled need is established and defined, brainstorming opens it up to a group of people to submit innovative and creative ideas related to how to solve this specific problem or to take advantage of the opportunity. This is done in a small group during a meeting, or it can be open to the entire organization to submit comments through the internal computer communication system. Ideas generated during the brainstorming session are then analyzed to select the best alternative.

- The remaining phases—During all the phases, problems arise and potential alternative situations occur where it is advisable to involve a group of people to share their ideas and get their buy-in to the end decision. Brainstorming is frequently used to collect alternative ideas at all of the phases.

HOW TO USE THE TOOL

Brainstorming or operational creativity combines a relaxed, informal approach to problem solving with lateral thinking. In most cases, brainstorming provides a free and open environment that encourages everyone to participate. While brainstorming can be effective, it is important to approach it with an open mind and a spirit of nonjudgment. Otherwise, people clam up, the number and quality of ideas plummets, and morale can suffer.

Brainstorming is perhaps the most widely recognized technique used to encourage creative thinking. It is also one of the most important tools any individual can have in his or her improvement arsenal. Problem-solving groups can take advantage of brainstorming techniques during several different phases of their operation.

Brainstorming is an intentionally uninhibited technique for generating the greatest possible number of ideas. Group members suggest as many ideas as they can about a given subject. The quantity of ideas is more important than their quality; each idea will be evaluated later. Groups can use this idea-generating technique to identify work-related problems, their causes, and possible solutions.

A brainstorming group exhibits the following characteristics:

- Has 4 to 12 members
- Determines the problem to be addressed
- Understands the problem
- Records all suggestions
- Gives each member the same opportunity to express opinions
- Encourages all ideas without criticism
- Has a leader who conducts the meeting, which includes keeping the group focused on the selected problem

Brainstorming uses the thinking resources of the entire problem-solving group. The ideas generated by a group are likely to be much more numerous and creative than those of an individual. The group generates a large number of ideas, accepts all of them, and writes them all down, even the silly and frivolous ones. At the end of the session, or later on in the problem-solving process, the group screens the ideas for the good ones.

Brainstorming is used to generate a large number of new and creative ideas. How can it help to determine the best solutions for important problems? Start with the assumption that people are conscientious, aware of problems affecting the quality of their work, and have thought about solutions. There are many good solutions incubating in the minds of people. Brainstorming is an opportunity for bringing these ideas out for consideration. This process helps meet the unending challenge of finding the right solutions to real problems. It recognizes that if certain conditions exist, people can participate in a creative process that is self-fulfilling, improves the quality of their work, and makes use of the organization's most valuable asset—the people in it.

Using Brainstorming

Groups use brainstorming when identifying and analyzing a problem and when looking for solutions. Constant attention must be given to the essentials of brainstorming such as

- Is everyone thinking about the same problem?
- Are all ideas encouraged and accepted without criticism?
- Are all ideas recorded?
- Do all of the group members have an equal chance to participate?

If all of these conditions are not satisfied, it is not a brainstorming session. Brainstorming is not a meeting in which everyone speaks at the same time, nor is it an unorganized *bull session*.

There are times when the group is faced with an unusual or difficult situation, one that cannot be solved through experience, formula, or some other known method. In these cases, a stronger technique is needed. Brainstorming facilitates diversity in thinking and the production of many ideas. It is not used to produce a single line of thought, nor is it the answer to a problem that has only one solution. Brainstorming is used when creative solutions are required. It is a cooperative, creative technique to be used when individual efforts do not yield satisfactory results.

PREPARING FOR A BRAINSTORMING SESSION

Brainstorming is worthwhile only when the problem to be solved can first be identified. All the members of the group should be aware of what the problem is, and they need to see all of the data relating to the present situation. Select a suitable meeting place for the occasion. The room needs to be just large enough to accommodate the group comfortably; too much room often leads to a loss of unity and makes it harder for the individual participants to coalesce into a group, a phenomenon critical to forming a collective intellect that operates as a single entity.

A relaxed atmosphere, which can readily accommodate laughter, is best for a productive and creative session. It allows the participants to verbalize their *offbeat* ideas by presenting them tongue-in-cheek. Experience has proven that often in the resulting laughter, a voice rises up to say, "You know, that's not such a crazy idea." Thus, the unusual thought often leads to a viable avenue or idea.

Brainstorming Techniques

There are several techniques that can be used to guide and expand brainstorming.

1. *Idea spurring.* The leader can ask questions like
   - Can we make these smaller?
   - What can we add?
   - What can we combine or package with this idea?

   The questions are designed to break down any mental barriers the group may have.
2. *Participation in sequence.* It is a good idea to ask for ideas to be contributed in turn, beginning with one person and going all the way around the group. This technique can be used at the beginning of a session to ensure that everyone participates, even shy members. If a person cannot think of anything to contribute, the appropriate response is *pass.* A good idea may occur before the next turn.
3. *Incubation.* Incubation is a process that may go on between brainstorming sessions. The initial brainstorming session gives the subconscious mind suggestions. The subconscious slowly works on

these suggestions and sometimes generates very creative ideas. In the incubation process, the members of the group just let the mind do its work between sessions.

Guidelines for Brainstorming (Harrington and Lomax, 2000)

- *Generate a large number of ideas.* Do not inhibit yourself or others; just let the ideas out. Say whatever comes into your mind and encourage others to do the same. The important thing is quantity.
- *Freewheeling is encouraged.* Even though an idea may be half-baked or silly, it has value. It may provoke thoughts from other members. Sometimes, making a silly suggestion can spur another idea you did not know you had.
- *Do not criticize.* This is the most important guideline. There will be ample time later to sift through the ideas for the good ones. During the session, you should not criticize ideas because you may inhibit other members. When you criticize the half-baked ideas, you throw away the building blocks for the great ones.
- *Encourage everyone to participate.* Everyone thinks and has ideas, so allow everyone to speak up. Speaking in turn helps; solicit ideas clockwise around the group. Encourage everyone to share his or her ideas.
- *Record all ideas.* Appoint a recorder to note everything suggested. The ideas should not be edited; rather, they should be jotted down just as they are mentioned. Keep a permanent record that can be read at future meetings. You may want to read through the list and take *inventory* a few times; this process sometimes stimulates more ideas.
- *Let ideas incubate.* Once you have started brainstorming, ideas will come more easily. You are freeing your subconscious mind to be creative. Let it do its work by giving it time. Do not discontinue your brainstorming sessions too soon; let some time go by to allow those ideas to develop by themselves.
- *Select an appropriate meeting place.* A casual place that is comfortable and the right size will greatly enhance a brainstorm session.

What Inhibits Good Brainstorming?

In as much as brainstorming is the most widely used tool, it is also the most misused. If the rules for brainstorming are not conscientiously followed, the quality of the session will suffer. A frequent mistake made by a session leader is to allow criticism of suggestions during brainstorming. This turns off other members of the group and keeps the group synergy from building. Brainstorming is not effective unless all members of the group participate. The leader should be especially careful to see that everyone is given the opportunity to contribute to the brainstorm list. It is the responsibility of the leader of the brainstorming group to encourage the members to comply with the rules for brainstorming.

Brainstorming helps us to release many subconscious creative ideas. Any problem-solving group may use brainstorming to identify problems in their work area. Brainstorming is an effective technique for directing the effort and attention of a group in a systematic problem-solving process.

Options: 3 General Ways That Brainstorming Can Be Conducted

There are at least three general ways that brainstorming can be conducted:

1. The brainstorming team agrees on the subject and takes some time personally to consider comments related to the subject. They then go around from one member to the next member of the brainstorming team, getting and recording the input from each member. This continues to go around the team, and if an individual has nothing to comment they just state *pass*, indicating they had no input. This continues until all the ideas have been recorded.
2. The brainstorming team agrees on the subject and takes personal time to write down their comments related to the subject on Post-its. When the brainstorming team has recorded their individual thoughts on the Post-its, they then go to the board where the facilitator is, and list some general breakouts related to the subject and paste their thoughts under the appropriate category.
3. When a subject that warrants conducting a brainstorming session is identified, this subject is posted on the Internet (company or general public Internet). The notice of the posting is transmitted to communities of interest, and anyone can add their thoughts on the subject. All thoughts are recorded and made available to anyone in the community of interest.

EXAMPLES

How should you manage a brainstorming session?

- Ask the group members to generate a large number of ideas.
- Record all ideas where they are visible to the entire group.
- Read through the list and restate the ideas several times throughout the session.
- Do not judge any of the ideas.
- Let ideas incubate.
- After all ideas have been recorded, review and clarify them.
- Determine what action needs to be taken.

List of Brainstorming Ideas

- Hire freight carrier based on lowest rate available
- Dock is overcrowded
- Historical trends of errors
- What is a *shipping error*?
- Time lag in order changes on computer
- When big customer push, switch labels and ship
- Shipping errors only in certain product lines
- Data entry complexity
- Difficult to measure true cost of errors
- High turnover among shippers
- New 11-digit code too long
- Bar codes damaged/unreadable
- Dock used to store material for return to vendors
- How many are paperwork errors—right product shipped?
- Shipping sometimes contracted directly by sales representative to rush orders
- Old shipping boxes easily damaged, requiring replacement
- Customer orders still initially handwritten
- How much does error cost?
- Wrong count by operators on production floor
- Need classification by type of error
- Frequently change freight carriers
- Too slow getting replacement product or paperwork
- Do not tell customer if shipping error is known but not detected by customer
- Allow changes to orders over the phone
- Sometimes substitute facsimile product is unavailable
- Computer system too slow—use handwritten forms instead

- Labels fall off boxes
- High turnover among data entry clerks
- No place to segregate customer returns
- Some new, reusable packaging has wrong bar codes
- Lack of training for data entry clerks
- Newest employees go to shipping
- Production bonus system encourages speed, not accuracy
- Dock is coldest place in winter, hottest place in summer
- How many customers lost that we do not know about?
- Certain clerks handle particular customers

SOFTWARE

Some commercial software available includes but is not limited to

- MindMap: www.novamind.com/
- Smartdraw: www.smartdraw.com/
- QI macros: http://www.qimacros.com

REFERENCE

Harrington, H.J. and Lomax, K. *Performance Improvement Methods.* New York: McGraw-Hill, 2000.

SUGGESTED ADDITIONAL READING

Asaka, T. and Ozeki, K., eds. *Handbook of Quality Tools: The Japanese Approach.* Portland, OR: Productivity Press, 1998.
Brassard, M. *The Memory Jogger Plus.* Milwaukee, WI: ASQ Quality Press, 1989.

15

Innovation Blueprint

Lisa Friedman

CONTENTS

All organizations are perfectly designed to get the results they get.

Arthur Jones

DEFINITION

An innovation blueprint is a visual map to the future that enables people within an enterprise or community to understand where they are headed and how they can build that future together. The blueprint is not a tool for individual innovators or teams to improve a specific product or service or to create new ones. Rather, an innovation blueprint is a tool for designing an enterprise that innovates extremely effectively on an ongoing basis.

USERS

The blueprint can be for an organization of any size; the numbers of people completing the blueprint can vary as well.

OFTEN USED IN THE FOLLOWING PHASES OF THE INNOVATION PROCESS

The following are the seven phases of the innovative cycle. An X after the phase name indicates that the tool/methodology can be used during that specific phase.

- Creation phase X
- Value proposition phase X
- Resourcing phase X
- Documentation phase X
- Production phase X
- Sales/delivery phase X
- Performance analysis phase X

TOOL ACTIVITY BY PHASE

An innovation blueprint is used before and during each of these phases, but is not a tool to be used *within* a particular phase per se. The innovation blueprint is a tool for building the kind of enterprise that can innovate effectively through each of these phases.

The blueprint maps the innovation strategy and best practices, organization design, tools, skills and teams, as well as the innovation leadership, stakeholder engagement, and support that need to exist in the background in order for each of these phases to be successful. It shows the *context and foundation* required for these phases to succeed, rather than tools or activities that occur within a phase.

HOW TO USE THE TOOL

Introduction

Innovation blueprints are tools for innovation leaders who want to ensure that their people have a shared view to their future innovation strategy, to where the enterprise is headed, and to what is needed to get there.

An innovation blueprint is a map that enables its users to see a shared view of the future, of where they want to go, and how they plan to get there (Figure 15.1). It clarifies three general categories of information that influences innovation capability:

- Where the world is going—whether the group in question goes there or not.
  - This helps clarify the scope of the market disruption in the external environment as well as significant new customer and market opportunities.

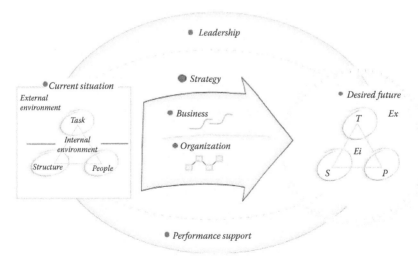

FIGURE 15.1
Innovation blueprint.

- Who they want to become in this emerging future.
  - *Who* includes the *what* of innovation strategy and best practices, but it also includes the values, identity, point of view, and culture required.
- How to enable their people to build this future, whether this includes 3 people, 30 people, 30,000, or hundreds of thousands.

While the blueprint can be for an enterprise of any size, the numbers of people completing the blueprint can vary as well.

- *Individuals.* Even for a very large enterprise, we have seen innovation blueprints completed by one single individual leader. Leaders often want to clarify their own understanding of where the enterprise needs to go to develop its innovation capabilities as a basis for setting their leadership direction.
- *Teams.* It is also quite common for leadership teams and department teams to create blueprints in groups of 5 to 25 people.
- *Large organizations.* At the larger end of the continuum, some organizations create *all hands* retreats or large-scale events of a hundred to several hundred people to ensure that many people see the same future direction and ambitions for their enterprise. If participants across an enterprise use the innovation blueprint as a tool to set their future direction together, they are more likely to be committed to implementation.
- *Linked blueprints at multiple levels within an organization or community.* The innovation blueprint is scalable, in that there may be a blueprint guiding a large company, supplemented with individual department blueprints as well. For example, a large media company may have its own blueprint that clarifies its goal of innovating for the emerging era of interactive trans-media. Each division (e.g., news, comedy, children's media, etc.) might then have their own innovation blueprints.

Innovation blueprints can also work across company boundaries; for example, a community might make an innovation blueprint for how it wants to deal with climate change, which might then lead to individual blueprints from various businesses, nonprofits, government agencies, or educational institutions. Or the government of a geographic region might create a blueprint for developing a center of expertise in smart and connected health, which might include separate but interlinked blueprints for local hospitals and health-care providers, technology companies, universities, start-ups, investors, or nonprofit patient advocacy groups. All groups might share the larger overarching blueprint together, while creating their own individual blueprints to specify the innovation strategy, practices, organization, skills, culture, leadership, stakeholder engagement, and support needed in their own organizations.

The blueprint is scalable, in that it can be used with an enterprise of any size: a team; a department; a company; or a larger network, community, or innovation ecosystem.

Why Use an Innovation Blueprint?

There are three fundamental reasons to use an innovation blueprint:

1. To create a *widely shared* view to the future that focuses innovation priorities.
   - Because the innovation blueprint provides a visual map, it helps people literally *be on the same page*. People innovating together can see a common vision of where they are headed and why. They can understand the kinds of innovation needed most, and how to collaborate effectively to create the intended results.
2. To promote *shared* innovation tools, architecture, and best practices.
   - When innovators throughout a large system have shared innovation methods, they can collaborate much more easily and quickly on innovation across boundaries.
3. To promote alignment in all the elements of the enterprise that contributes to innovation capability.
   - A blueprint can help ensure the organization as a whole is truly designed to deliver innovation, where all functions of the enterprise support this capability (e.g., overall organization design as well as individual functions such as finance, human resources [HR], information technology, or marketing). For example, if employees in a company are encouraged to innovate but their performance reviews never address this, or if they do not have a clear innovation process that helps them know how to progress their ideas, or if they cannot get time or budget for innovation, employees get the message that innovation is not really valued.

On the other hand, when the major functions in an organization are all aligned to deliver innovation, the overall result is a spirit and capability for *innovation everywhere.*

Just as the innovation blueprint is a tool that can be used with various sizes and scopes of enterprises, it is also a tool that can be completed in various levels of detail. It is similar to a business plan in this way. Just as a business plan may have an executive summary of a just a few pages or can be a very large document with supporting research and detail, an innovation blueprint can be used at any depth users choose. Each enterprise can determine the level of detail that works best for its current need. What is most important is that each major element is addressed. Providing a template for aligning key elements of the enterprise to deliver innovation is what makes the blueprint a powerful tool.

Elements of the Innovation Blueprint

Innovation in an enterprise typically needs to occur in both its *business* (its connection to customers and market, and in its products and services), and its *organization* (its organization structure; people's skills, talents, and ways of working together; and culture and climate). When business needs change or new opportunities arise in the marketplace, the organization needs to change as well in order to enable business success under the new conditions. On the blueprint in Figure 15.1, the top half of the map represents the business and the lower half represents the organization.

Business

The key elements that define the business of an enterprise are the external environment and the core task of the business.

- *External environment.* An enterprise exists within a given external environment of multiple forces: customers; competitors; strategic partners; cultural, economic, and social trends; regulations; science and technology; resources; and new market opportunities.
- *Task.* Task includes how the business generates customer value through its products and services, as well as its business model—how revenue is generated.

Organization

Organization includes the elements that can be arranged to help deliver business success: the organization's structure, its people, and its internal environment or culture.

- *Structure.* Organization structure can be thought of as riverbanks that guide how work flows within the enterprise. Structure can include organization design (divisions and departments, job descriptions, reporting relationships, roles, and responsibilities); locations and facilities; finance systems; information systems and tools; communication systems; and HR policies and practices—especially including performance management systems, incentives, and rewards.
- *People.* This category includes the kinds of people needed and how they work together, for example, skills and talents, demographics, teamwork, and the quality of communication and collaboration.
- *Internal environment.* This includes how it feels to work inside the enterprise or the organization's culture. It can include vision and values; sense of identity; emotional climate; and levels of trust, spirit, or energy.

STEP

The acronym STEP—for structure, task, environments (both internal and external), and people—helps frame these elements as one interconnected system (see Figure 15.2).

STEP and the Innovation Blueprint

The innovation blueprint looks at how well each element of the enterprise is positioned to support innovation. Elements of each STEP category are examined to see how well they currently support innovation and what is needed. While the questions can vary for each individual enterprise, typical questions include

- *External environment*
  - How is new technology influencing our industry? What are the risks and new opportunities this presents?

STEP

Enterprise
assessment

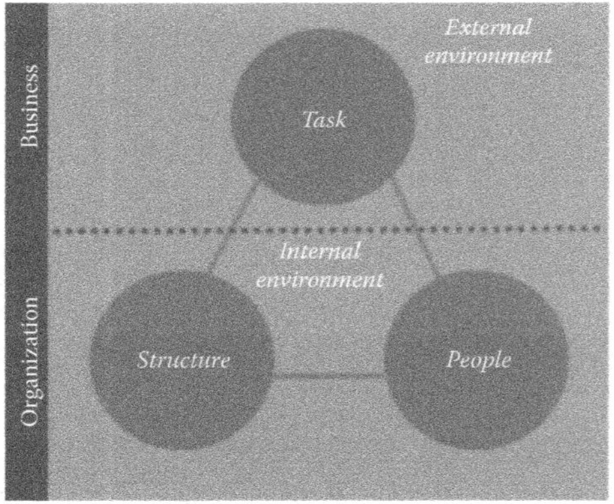

External environment
- Consumers
- Economy/market forces
- Competitors
- Suppliers
- Natural environment
- Society/community
- Science/technology
- Regulations
- Headquarters/parent organization

Task
- Solutions, products and services
- Standards
- Priorities and goals
- Core competencies
- Competitive advantage
- Skills
- Work flow

Internal environment
- Culture/climate
- Norms
- Values
- Mission
- Strategic intent
- Vision
- Positioning

Structure
- Roles and reporting relationships
- Reward systems
- Accountability/authority/status
- Policies and procedures
- Decision making and planning
- Reward facilities
- Resource management
- Communication

People
- Technical and management talent
- Skills mix
- Diversity
- Team dynamics/quality of relationships
- Cognitive and emotional experience/"maturity"
- Needs and expectations

FIGURE 15.2
STEP elements of the innovation blueprint.

- Are there significant changes in the economy, in regulations, in the environment, or in culture that affect our business?
- How are customer interests and needs changing?
- What new competitors are entering our market?
- Are there potential new partners for us in our industry?
- Is there a major transformation occurring in our industry? If so, do our people all understand this?
- *Task.* Given what we see in the external environment
    - What is our overarching innovation strategy? How do we envision our next generation of products and services?
    - What improvements do we need in our existing products and services?
    - Do we need to innovate our business model?
    - Do we have a clear and widely understood innovation system— with shared concepts, language, practices, and tools? Does everyone in our enterprise understand the changes driving our industry; how to generate new ideas; how to turn ideas into value propositions and how to collaborate with others to make these as strong as they can be; how to get funding, approval, and resources for innovation; and how to move ideas through to implementation?
    - Do we have a balanced innovation portfolio?
- *Structure*
    - Is our organization designed to enable innovation?
    - Do our facilities support innovation?
    - Do our HR policies and practices support innovation?
    - Does finance support innovation? Do we have a budget for innovation?
    - Does our legal department enable innovation rather than suppress it?
    - Do we have the information systems and tools to enable innovation? Do our people get the information they need, and can they collaborate with others?
        - For example, do we have tools to support creativity and collaboration, that is, knowledge management systems? Online innovation management platforms? Online learning applications?
- *People*
    - What skills and talents do our people need to innovate effectively?
    - Do our people have the right mix of skills and talents to create the next generation of products and services in our industry?

- Can our people collaborate across boundaries?
- Do our people regularly come up with new ideas, solutions to problems, and new ways to create customer value?

- *Internal environment*
  - Do we have a culture of innovation—a culture of creativity and experimentation?
  - Are our people energized and inspired about future opportunities?
  - Can anyone innovate? Do we have a culture of *innovation everywhere*?
  - Do our people trust, respect, and support each other?
  - Do our people want to be the best?

- *Innovation leadership.* Do our leaders
  - Treat innovation as a strategic priority?
  - Ensure the systems, practices, tools, events, and support are in place to enable innovation capability throughout the enterprise?
  - Role model innovation?
  - Lead and sponsor visionary innovation campaigns?
  - Participate on innovation panels and juries, to select projects to move forward?
  - Celebrate and reward inspiring example of innovation, and give genuine appreciation for learning from failure?
  - Coach and mentor innovators?

- *Stakeholder engagement*
  - Are people actively involved in developing the vision for the future of the enterprise, and are they committed to the aspirations associated with that vision?
  - Are people involved in improving current products, services, and processes, as well as creating breakthrough innovations?
  - Do people understand how to innovate in our organization?
  - Do we have events or online innovation platforms that give them a way to participate in strategic innovation campaigns?
  - How many ideas are innovators submitting?
  - How many people are engaged in commenting on improving others' ideas?
  - Are people satisfied with their ability to innovate in their day-to-day work and to participate in targeted innovation activities?

- *Innovation support.* Do we have
  - A team to build and monitor our innovation system?
  - Innovation training as needed—for both innovators and leaders?

- A way to monitor and measure innovation results?
- A way to stay in touch with the emotional climate around innovation?
- A way to continuously refine, and even innovate, our innovation system?

Current and *Future* Elements of the Blueprint

The innovation blueprint shows both current STEP and future STEP (see Figure 15.3). Having both the current innovation capability and the capabilities needed for the future in a single map helps clarify the requirements for moving to the intended future.

Three Drivers of Innovation

The STEP components in the blueprint illustrate the plan for building an enterprise with strong innovation capabilities, but it is people who turn plans into action. In particular, there are three groups of people who determine if innovation capability truly comes alive in an organization:

- *Innovation leaders.* Innovation leaders inspire, communicate, authorize, fund, and enable stakeholders to build the future. Leaders can be formal or informal, and can be internal or external. Sometimes, community or government leaders, or leaders from partners, suppliers, or customer groups might be needed for success as well.
- *Key stakeholders.* Leaders lead the enterprise to the envisioned future, but it is generally the stakeholders who actually build the

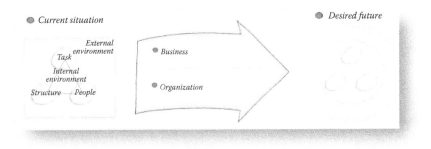

FIGURE 15.3
Current and future in the innovation blueprint.

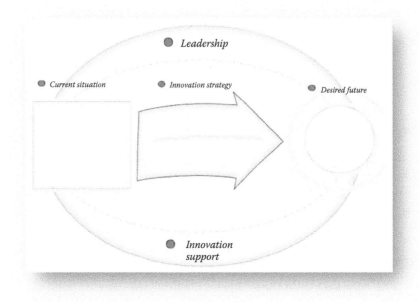

FIGURE 15.4
Three drivers of innovation: leadership, stakeholders, and innovation support.

future. This group typically includes employees, partners, suppliers, and sometimes even customers.

- *Innovation support.* Innovation support is most often a team that coordinates innovation activities and ensures that tools, training, and connection to expert resources are in place as needed. If leaders inspire and authorize stakeholders to build the future, then the innovation support team ensures that stakeholders are capable to build the future, and that the process is coordinated and stays on track.

The arches and the arrow in the blueprint (see Figure 15.4) show the three groups of people who determine whether the enterprise innovates to truly move from its current state to its intended future.

EXAMPLES

How do you gather the information for the blueprint? The most common methods for gathering the information for the blueprint include

- *Individual interviews.* Beginning with individual interviews to gain diverse perspectives on each of the elements of the blueprint.
- *Survey.* Conducting a survey across a wider population throughout the enterprise.
- *In-person events with a single group blueprint.* Participants may meet in retreats or workshops lasting from one day to several days, in order to gain group agreement about a blueprint. This is the most powerful use of the tool, as the process of completing the innovation blueprint as a group promotes shared innovation strategy, clarity about the organization design required, commitment to innovation leadership and support, and shared agreement about critical success factors and next steps (see Figure 15.5).
  - In these events, a large paper copy of a blank blueprint is typically put on the wall. A blueprint can be printed or can be blank paper with the blueprint drawn by hand.
  - Common ways to add content to the blueprint include
    - Participants in the session call out feedback for a facilitator to write into each section.
    - Participants or small groups add their information to Post-it notes and then describe their input to the group as the note is added to the map.
    - Small groups get cardboard segments to add their information to each section of the blueprint (e.g., they can get pie-shaped wedges to fill in their sections on the part of the blueprint that are circles). Each group then describes their input to the large group as they add their piece to the blueprint (see Figure 15.6).

FIGURE 15.5
Workshop participants working on a blueprint.

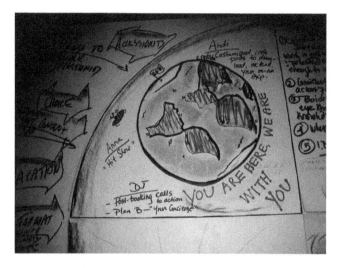

FIGURE 15.6
Cardboard segment for a small group's addition to a large group blueprint.

- *In-person events with multiple individual blueprints.* In these sessions, participants create a very high-level map for the whole enterprise as a large group—in an hour or two—and then individuals or small groups make more detailed personal or team-sized blueprints for their own division, department, or team (see Figure 15.7, right photo). If there are a large number of individuals, the larger group may then take a *gallery walk*, where individuals or teams volunteer to give the group a quick overview of their maps (see Figure 15.7).

FIGURE 15.7
Participants working on innovation blueprints.

SOFTWARE

- ValueSTEP is an online survey that assesses innovation capability in each of the STEP categories (see "STEP" section).
- ValueSTEP also gathers information about the three drivers for innovation: innovation leadership, stakeholder engagement, and innovation support (see discussion in "STEP" section).

More information can be found at www.enterprisedevelop.com in the Tools section.

SUGGESTED ADDITIONAL READING

Friedman, L. and Gyorffy, L. *Leading Innovation: Ten Essential Roles for Harnessing the Creative Talent of Your Enterprise*. In: P. Gupta and B. Trusko, eds. *The Global Handbook of Innovation Science*. New York: McGraw-Hill Education, 2014.

Friedman, L. and Gyr, H. *Creating Your Innovation Blueprint: Assessing Current Capabilities and Building a Roadmap to the Future*. In: P. Gupta and B. Trusko, eds. *The Global Handbook of Innovation Science*. New York: McGraw-Hill Education, 2014.

Friedman, L. and Gyr, H. *The Dynamic Enterprise: Tools for Turning Chaos into Strategy and Strategy into Action*. San Francisco: Jossey-Bass Business and Management Series, John Wiley, 1998.

Gyorffy, L. and Friedman, L. *Creating Value with CO-STAR: An Innovation Tool for Perfecting and Pitching Your Brilliant Idea*. Palo Alto, CA: Enterprise Development Group Inc. Publishing, 2012.

16

Lead User Analysis

Charles Mignosa

CONTENTS

DEFINITION

Lead users are users of a product or service who provide input to the organization related to new products and services because they foresee needs that are still unknown to the marketplace. Lead users innovate, and therefore are considered to be part of the creative consumers' phenomenon, that is, those "customers who adapt, modify, or transform a proprietary offering" (Berthon et al., 2007).

USER

This tool can be used by individuals, but its best use is with a group of four to eight people. Cross-functional teams usually yield the best results from this activity.

OFTEN USED IN THE FOLLOWING PHASES OF THE INNOVATIVE PROCESS

The following are the seven phases of the innovative cycle. An X after the phase name indicates that the tool/methodology is used during that specific phase.

- Creation phase X
- Value proposition phase
- Resourcing phase
- Documentation phase
- Production phase
- Sales/delivery phase
- Performance analysis phase

TOOL ACTIVITY BY PHASE

- Creation phase—In this phase, the lead users are provided with a goal to achieve and in many cases the means to achieve it.

HOW THE TOOL IS USED

Introduction

Lead users are

- Users who have real-world experience and can provide accurate data regarding it.
- When needs evolve rapidly, only users at the front of the trend will have insight into tomorrow's requirements.
- Studies of industrial product and process innovations (Mansfield, 1968) show that those who expect a high benefit from a solution to a need tend to experiment with solutions on their own, and so can provide the richest need and solution data to market researchers.

A lead user market research study involves four major steps, which are spelled out in detail as follows (von Hippel, 1986, 1988).

1. Start of the lead user process:
   a. Building an interdisciplinary team
   b. Defining the target market
   c. Defining the goals of the lead user involvement
2. Identification of needs and trends
   a. Interviews with experts (market technology)
   b. Scanning of literature, Internet, databanks
   c. Selection of most attractive trends
3. Identification of lead users
   a. Networking-based search for lead users
   b. Investigation of analogous markets
   c. Screening of first ideas and solutions generated by lead users
4. Concept design
   a. Workshop with lead users to generate or to improve product concepts
   b. Evaluation and documentation of the concepts

Techniques commonly used include but are not limited to

- Delphi
- Trend extrapolation
- Interview with experts

Basic Steps in Managing a Lead User Group

The process has two selection steps yielding a subset of lead users representing lead users in each subcategory related to the product of service (see Figure 16.1).

1. Overview from publications, Internet, and related lists.
   a. Compile the results into related categories.
   b. Identify a subset of experts' leading players in each category.
2. From the leading players in the fields, identify a second subset of lead users.
3. Identify another subset of lead users in each category who fulfill the lead user criteria with solution experience and an interest in proved solutions.

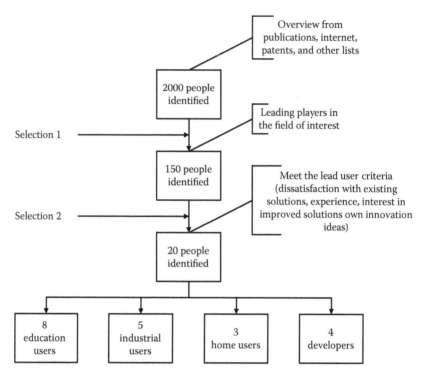

FIGURE 16.1
Lead user analysis.

EXAMPLE

An example is included in the section of this chapter entitled "How the Tool Is Used."

SOFTWARE

Some commercial software available includes but is not limited to

- Delphi XE7: Embarcadero.com
- Graphics programs
  - MiniTab
  - Excel
  - QI macros

REFERENCES

Berthon, P.R., Pitt, L.F., McCarthy, I., Kates, S.M. When customers get clever: Managerial approaches to dealing with creative consumers. *Business Horizons*, vol. 50, no. 1, pp. 39–47, 2007. doi: 10.1016/j.bushor.2006.05.005.

Mansfield, E. *Industrial Research and Technological Innovation: An Econometric Analysis.* New York: W.W. Norton & Company, 1968.

von Hippel, E. Lead users: A source of novel product concepts. *Management Science*, vol. 32, no. 7, pp. 791–805, July 1986.

von Hippel, E. *The Sources of Innovation.* New York: Oxford University Press, 1988.

17

Lotus Blossom

Frank Voehl

CONTENTS

DEFINITION

Lotus blossom is a technique based on the use of analytical capacities and helps generate a great number of ideas that will possibly provide the best solution to the problem to be addressed by the group. It uses a six-step process.

USER

This tool can be used by individuals or groups, but its best use is with a group of four to eight people. Cross-functional teams usually yield the best results from this activity.

OFTEN USED IN THE FOLLOWING PHASES OF THE INNOVATIVE PROCESS

The following are the seven phases of the innovative cycle. An X after the phase name indicates that the tool/methodology is used during that specific phase.

- Creation phase X
- Value proposition phase X
- Resourcing phase
- Documentation phase
- Production phase X
- Sales/delivery phase X
- Performance analysis phase

TOOL ACTIVITY BY PHASE

Lotus blossom is a tool that can be effectively used whenever an individual or group is having problems coming up with a creative solution to a problem or a new product or service. It is particularly effectively used during the creation, production, sales, and delivery phases.

Lotus blossom is a creative–innovative–thinking technique that will help you expand your thinking beyond the usual paths of thinking and its boundaries. Invented in Japan by Yasuo Matsumura, the lotus blossom technique adds focus and power to the classic creativity technique of brainstorming and starbursting. Once mastered, lotus blossom helps you innovate and create more and higher-quality ideas for products and services. It helps individuals find new and unusual ways of improving their products and processes, and helps them solve a variety of problems they frequently encounter. According to popular innovation author Michael Michalko, lotus blossom helps one to organize their thinking around significant themes, helping them explore a number of alternate possibilities and ideas.

When to Use Lotus Blossom

1. Use it when you want to develop creative ideas.
2. Use it when you are having problems creating more ideas.
3. Use it when you are trapped by a single mode of thinking.
4. Use it to create seeds of ideas that can trigger further good ideas.
5. Use it in combination with other creativity tools, such as starbursting and brainstorming.

Background

The International Association of Innovation Professionals and other research models have shown that while we were all born as spontaneous, creative thinkers, a great deal of our education may be regarded as the inculcation of mind-sets. In grammar school, we were taught how to handle problems and new phenomena with fixed mental attitudes based on what past thinkers thought, which, in many cases, predetermine our response to problems or situations. Typically, we think on the basis of similar problems encountered in the past. When confronted with problems, we fixate on something in our past that has worked before. Then, we analytically select the most promising approach based on past experiences, excluding all other approaches, and work within a clearly defined direction toward the solution of the problem.

> Once we think we know what works or can be done, it becomes hard for us to consider alternative ideas. We tend to develop narrow ideas and stick with them until proven wrong. Following is an interesting experiment, which was originally conducted by the British psychologist Peter Watson, which demonstrates the way we typically process information. Watson would present subjects with the following three numbers in sequence: 2... 4 ... 6 .... He would then ask subjects to explain the number rule for the sequence and to give other examples of the rule. The subjects could ask as many questions as they wished without penalty. He found that almost invariably most people will initially say, "4, 6, 8" or some similar sequence.

And Watson would say, yes, that is an example of a number rule. Then they will say, "20, 22, 24" or "50, 52, 54," and so on—all numbers increasing by two. After a few tries, and getting affirmative answers each time, they are confident that the rule is numbers increasing by two without exploring alternative possibilities.

Actually, the rule Watson was looking for is much simpler—it is simply numbers increasing. They could be 1, 2, 3 or 10, 20, 40 or 400, 678, 10,944. And testing such an alternative would be easy. All the subjects had to say was 1, 2, 3 to Watson to test it, and it would be affirmed. Or, for example, a subject could throw out any series of numbers, for example, 5, 4, 3, to see if they got a positive or negative answer. And that information would tell them a lot about whether their guess about the rule is true.

The profound discovery Watson made was that most people process the same information over and over until proven wrong, without searching for alternatives, even when there is no penalty for asking questions that give them a negative answer. In his hundreds of experiments, he, incredibly, never had an instance in which someone spontaneously offered alternative hypotheses to find out if it were true. In short, his subjects did not even try to find out if there is a simpler or even, another, rule.

The lotus blossom technique focuses the power of brainstorming on areas of interest. It does so through the use of a visual representation of ideas and is similar to a mind map, but is more structured and pushes you in ways you do not experience in classic mind mapping. We usually start with a central idea or theme, and then expand our thinking outward with solution areas or related themes in an iterative manner. The technique encourages you to have a fully fleshed out idea space before you can consider it complete.

Using the Tool

1. *Start by describing the problem:* Write the central problem in the center of the diagram.

    Start with a description of the problem you are facing. Write it on a card or Post-it note and put it in the middle of a large working area. If you are working in a group, this works well with a vertical work area, such as a wall or large pinboard. You can also use the floor or a large table top.

2. *Surround it with ideas*: Use other tools for creating ideas to create a set of ideas on how to solve the problem.

These should be as different from one another as possible. Write each idea on a card or Post-it note of its own and place it around the problem description. Eight ideas fit neatly, as below. You can also do six, in a hexagonal shape. In placing the ideas around the original problem, you can put them down in a *knight's pattern*. This helps mix up the ideas and generates more different thinking. When you are working in a group of people, engage everyone. Perhaps they could each create a different idea.

3. *Unfold the lotus blossom*

Make a copy of each of the idea cards and place them radially further out from the cluster (the stamen and pistil of the flower) that you created in the previous steps. Now repeat the previous step of the process, surrounding each of the copied idea cards with secondary ideas, using only the copied idea cards as stimuli. This should result in ideas that are further removed from the original problem. This can lead to many ideas, as in Figure 17.1. You do not have to fill in every space—if ideas run out, you can move on. Also, if an idea seems to be leading somewhere, you can repeat the whole process ad infinitum until you get somewhere or nowhere.

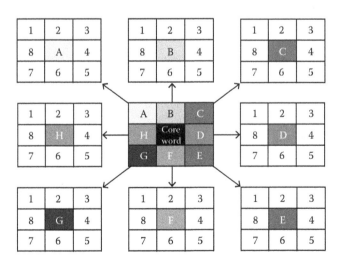

FIGURE 17.1
Typical lotus blossom layout.

When you are working in a group of people, you can rotate them around the lotus. Thus, a person puts one secondary idea against a copied idea and then moves on to the next copied idea. This creates a dance around the ideas, moving the thinking on at each step.

4. Continue the process until the lotus blossom diagram is completed (see Figure 17.1).

Advantages

- Helps you write the significant themes, components, or dimensions of your subject in a concise manner, using the surrounding circles (labeled A to H) surrounding a central and visible theme.
- Lists the optimal number of themes for a manageable diagram (between six and eight), and gets it on one page. If you have more than eight, you will need to make additional diagrams, making this a disadvantage as well.
- Helps you to ask questions like, What are my specific innovation objectives? What are the constants in my innovation problem? If my subject were a book, what would the chapter headings be? What are the dimensions of my problem?
- Uses the ideas written in the circles as the central themes for the surrounding lotus blossom petals or boxes. Thus, the idea or application you wrote in circle A would become the central theme for the lower middle box A.
- The outer blossoms follow the same pattern as the inner blossom, making it easy to deploy. Each contains an inner box surrounded by outer boxes that are related in some way, either a concept, idea, or solution.

Disadvantages

- Creative geniuses do not always think this way. The creative genius will always look for a multiplicity of ways to approach a subject. It is this willingness to entertain different perspectives and alternative approaches that broadens their thinking and opens them up to new information and the new possibilities that the rest of us do not see.
- When Charles Darwin first set to solve the problem of evolution, he did not analytically settle on the most promising approach to natural selection and then processed the information in a way that

would exclude all other approaches. Instead, he initially organized his thinking around significant themes, principally eight, of the problem, which gave his thinking some order but with the themes connected loosely enough so that he could easily alter them singly or in groups.

- The genius themes help capture thoughts about evolutionary change by allowing us to reach out in many alternative directions at once and pulling seemingly unrelated information into a coalescent body of thought.

EXAMPLE

Example 1: Adding Value to a Website

Figure 17.2 was taken from http://www.innovationmanagement.se/imtool -articles/creative-thinking-technique-lotus-blossom/.

Example 2: Adding Value to Your Organization

Suppose, for example, you want to create more value for your organization by increasing productivity or decreasing costs. You would write "Add Value" in the center box. Next, write the eight most significant areas in your organization where you can increase productivity or decrease costs in the circles labeled A to H that surround your central box. Also write the same significant areas in the circles with

Improve look/feel	Improve speed	Offer web coupons
Cross-site promotions	Website	Advertise the site
Have signs around store	Allow people to order from web pickup in store	Allow return of web purchases to store

FIGURE 17.2
Typical example of a lotus blossom layout.

the corresponding letters spread around the diagram. In my example, I selected the themes *suppliers, travel expenses, partnerships, delivery methods, personnel, technology, facilities,* and *evaluation.* Also write the same significant areas in the circles with the corresponding letters spread around the diagram. For instance, in the sample diagram, the word *technology* in the circle labeled A serves as the theme for the lower middle group of boxes. Each area now represents a theme that ties together the surrounding boxes.

For each theme, try to think of eight ways to add value. Phrase each theme as a question to yourself. For example, ask, "In what ways might we use technology to increase productivity?" and "In what ways might we use technology to decrease expenses?" Write the ideas and applications in the boxes numbered 1 through 8 surrounding the technology theme. Do this for each theme. Think of eight ideas or ways to make personnel more productive or ways to decrease personnel expenses; eight ideas or ways to create more value for your delivery methods, your facilities, and so on. If you complete the entire diagram, you will have 64 new ideas or ways to increase productivity or decrease expenses.

When you write your ideas in the diagram, you will discover that ideas continually evolve into other ideas and applications, as ideas seem to flow outward with a conceptual momentum all their own.

An important aspect of this technique is that it shifts you from reacting to a *static* snapshot of the problem and will encourage you to examine the significant themes of the problem and the relationships and connections between them. Sometimes when you complete a diagram with ideas and applications for each theme, a property or feature not previously seen will emerge. Generally, higher-level properties are regarded as emergent—a car, for example, is an emergent property of the interconnected parts.

If a car were disassembled and all the parts were thrown into a heap, the property disappears. If you placed the parts in piles according to function, you begin to see a pattern and make connections between the piles that may inspire you to imagine the emergent property—the car, which you can then build. Similarly, when you diagram your problem thematically with ideas and applications, it enhances your opportunity to see patterns and make connections. The connections you make between the themes and ideas and applications will sometimes create an emergent new property or feature not previously considered.

Lotus Blossom Case Study #1: Darwin

Charles Darwin used his themes to work through many points that led to his theory of evolution by helping him comprehend what is known and to guide in the search for what is not yet known. He used them as a way of classifying the relation of different species to each other, as a way to represent the accident of life, the irregularity of nature, the explosiveness of growth, and of the necessity to keep the number of species constant. Over time, he rejected some of his themes—the idea of direct adaptation, for instance. Some were emphasized—the idea of continuity. Some were confirmed for the first time—the idea that change is continuous. Some were recognized—the frequency of variation. By adjusting and altering the number of themes and connections, Darwin was able to keep his thought fluid and to bring about adaptive shifts in his thinking. He played the critic, surveying his own positions; the inventor, devising new solutions and ideas; and the learner, accumulating new facts not prominent before.

The key for innovation is that by organizing his thinking around loosely connected themes, Darwin expanded his thinking by inventing alternative possibilities and explanations that otherwise may have been missed, at least until another time.

Lotus Blossom Case Study #2: Scenario Planning

This technique can also be used in scenario planning and is very useful for forecasting strategic scenarios. It is designed for groups and is used to provide a more in-depth look at various solutions to problems. It begins with a central core idea surrounded by eight empty boxes or circles. Using brainstorming, eight additional ideas (solutions or issues) are written in these boxes. In the next step, each of these eight ideas becomes the core of another set of eight surrounding empty boxes, which are filled in by new ideas using brainstorming. The process continues until a satisfactory solution or a sufficient number of ideas have emerged (Higgins, 1996).

SOFTWARE

No particular software is recommended.

SUGGESTED ADDITIONAL READING

CreatingMinds.org. Lotus blossom. Available at http://creatingminds.org/tools/lotus_blossom .htm.

Davila, S. Creative thinking lotus blossom: Where are you going? Sample with complete lesson plan designed to supplement as a reference. New York: Kenyon. Available at http://saradavila.com/front/2014/03/08/creative-thinking-lotus-blossom-where-are -you-going/.

Higgins, J. *101 Creative Problem Solving Techniques: The Handbook of New Ideas for Business.* Revised edition. Winter Park, FL: New Management Pub Co., 2005.

Kenyon, D.M. *The Lotus Blossom: A Novel About Waking Up.* New York: Thought Locker Media, 2012.

Mauzy, J. and Harriman, R. *Creativity Inc.: Building an Inventive Organization.* Boston: Harvard Business School, 2003.

Nolan, V. *The Innovators Handbook.* New York: Penguin, 1989.

Prince, G.M. *The Practice of Creativity: A Manual for Dynamic Group Problem-Solving.* Battleboro, VT: Echo Point Books & Media, LLC, 2012. 0-9638-7848-4.

18

Matrix Diagram/Decision Matrix

H. James Harrington

CONTENTS

We frequently need a way to select the right alternative.

H. James Harrington

DEFINITION

Matrix diagram, also called decision matrix, is a systematic way of selecting from larger lists of alternatives. They can be used to select a problem from a list of potential problems, select primary root causes from a larger list, or to select a solution from a list of alternatives.

USER

This tool can be used by individuals, but its best use is with a group of four to eight people. Cross-functional teams usually yield the best results from this activity.

OFTEN USED IN THE FOLLOWING PHASES OF THE INNOVATIVE PROCESS

The following are the seven phases of the innovative cycle. An X after the phase name indicates that the tool/methodology is used during that specific phase.

- Creation phase X
- Value proposition phase
- Resourcing phase X
- Documentation phase
- Production phase
- Sales/delivery phase X
- Performance analysis phase

TOOL ACTIVITY BY PHASE

- Creation phase—Frequently during this phase, there are a number of alternative potential approaches to addressing an opportunity or problem. In these cases, the decision matrix provides insight into the one way to which alternative should be pursued first.
- Financing phase—During the financing phase, there are often a number of different financing alternatives that need to be evaluated. The decision matrix helps the group come to common agreement on prioritizing the financing alternatives.
- Sales/delivery phase—There are often many different marketing and sales approaches that need to be considered before the final delivery

process is put in place. Alternatives like, Should we market ourselves? Should we set up our own stores? Should we run through major wholesalers? Do we want to get into the retail business? Do we want to use outside marketing representatives? Do we need different marketing approaches for Europe? The decision matrix provides an excellent way to help the sales team define the options they would like to implement.

HOW TO USE THE TOOL

The matrix diagram is an approach that assists in the investigation of relationships. While there are many variations of matrix diagrams, the most commonly used is the decision, or prioritization, matrix. These come in two basic formats—the *L-shaped matrix* and the *T-shaped matrix.*

L-Shaped Matrix (Harrington and Lomax, 2000)

We will start by showing in Figure 18.1 a relatively simple L-shaped matrix that compares two sets of information.

As you can see from the example, we are comparing several automobile dealers (our choices) with a predetermined set of decision criteria. Now all that remains is to determine which type of ranking method we will use. There are four basic types of ranking methods:

1. *Forced choice*: Each alternative is ranked in relation to the others. The alternative best meeting the criteria gets a score equal to the number of alternatives. Since we have five dealers in our example, the worst would get a 1 and the best a 5.
2. *Rating scale*: Each alternative is rated independently against an objective standard. For example, a 1–10 scale would have 1 = very low (does not meet the standard at all) and 10 = perfect (absolutely meets the standard).
3. *Objective data*: Here, we enter actual data, rather than the opinions of the individual(s) doing the ranking.
4. *Yes/No*: If the criteria are expressed in absolute terms, so an alternative either meets the criteria or not, a Y for yes, or an N for no may be entered to indicate conformance or nonconformance.

Criteria Choice	Recommended by friends	Good selection of cars	Good service department	Free loaner cars	Free drop-off and pick-up
Dealer 1					
Dealer 2					
Dealer 3					
Dealer 4					
Dealer 5					

FIGURE 18.1
Example of an L-shaped matrix.

Figure 18.2 shows our automobile dealership example using the simple Yes/No ranking criteria.

As you can see, the easiest ranking method—the Yes/No approach—often leaves the user with little information in which to make a decision. In Figure 18.2, all five dealers got three yes.

Let us try ranking our choices using the forced choice method (see Table 18.1). Remember, this method ranks each alternative in relation to the others. In this case, the dealer meeting our criteria the best will get a 5 and the worst will get a 1 (since we have five choices).

Criteria Choice	Recommended by friends	Good selection of cars	Good service department	Free loaner cars	Free drop-off and pick-up
Dealer 1	(Y)	(Y)	N	N	(Y)
Dealer 2	(Y)	N	(Y)	(Y)	N
Dealer 3	N	(Y)	(Y)	N	(Y)
Dealer 4	(Y)	(Y)	(Y)	N	N
Dealer 5	(Y)	N	(Y)	(Y)	N

FIGURE 18.2
Example of an L-shaped matrix using the Yes/No ranking method.

TABLE 18.1

Example of an L-Shaped Matrix Using the Forced Choice Ranking Method

Criteria Choice	Recommended by Friends	Good Selection of Cars	Good Service Department	Free Loaner Cars	Free Drop-Off and Pick-Up	Totals
Dealer 1	5	4	1	2	5	17
Dealer 2	3	1	2	5	3	14
Dealer 3	1	3	3	1	4	12
Dealer 4	4	5	5	3	1	(18)
Dealer 5	2	2	4	4	2	14

Now we have information that might allow us to make a decision. As you can see, dealer 4 scored the highest with 18. Does this mean you should automatically buy your car from him? Not necessarily. Although dealer 4 did score the highest overall, he scored the lowest on *free drop-off and pick-up*. If this were a critical element to the potential buyer, he or she might want to consider the second choice, dealer 1.

This is where using the objective data method might be of assistance. The person or group doing the ranking might even consider using a combination of ranking methods. This is certainly an option, but it makes the final selection a bit more complex.

T-Shaped Matrix

The second format we mentioned was the T-shaped matrix. While the L-shaped matrix compares two sets of information, the T-shaped matrix compares two sets of information to a third. An example of this could be a corporation's training program. We could compare the type of training available with departments that need the training and training providers. Figure 18.3 shows an example of the T-shaped matrix format.

There are many approaches to designing and developing a matrix diagram. Listed below are five steps you may find useful in developing a matrix diagram that is just right for your purpose.

Step 1. Determine the task. Are we looking at two elements or three? What should the desired outcome look like? Is the matrix to be used as a problem-solving tool or a planning graph? Is it a *stand-alone* tool that leads us to action or will we use it in conjunction with other tools, such as a tree diagram or relation diagram?

Step 2. Select the matrix format. If we are reviewing the relationships of two elements, you may want to use the L-shaped matrix. If you add a third element, you will want to use the T-shaped matrix.

Step 3. Determine the criteria for evaluating alternatives. A typical list of criteria is presented below:
- Customer impact
- Number of customers affected
- Within control of the team
- Within influence of the team
- Cost of quality

Training providers	OD dept.			X			X
	Quality group	X			X	X	
	Direct supervisor						
	Outside resource	X	X	X			
Training available		**TQM tools**	**Computer skills**	**Specific job skills**	**Team building**	**Effective meetings**	**Leadership skills**
Departments requiring specialty training	Engineering	X			X		
	Manufacturing	X			X		X
	Finance	X	X	X			
	Sales	X			X	X	
	Marketing	X			X		

FIGURE 18.3
Example of a T-shaped matrix.

- Rework
- Frequency of occurrence
- Cycle time impact
- Revenue impact
- Return on investment
- Complexity of analysis
- Time to develop a solution
- Durability of solution
- Cost to implement solution
- Availability of measurements
- Etc.

The criteria should be worded in terms of the ideal result, not be worded neutrally. For example, a criterion could be *Easy to implement*, but not *Ease of implementation*.

Step 4. Determine the weights for the individual criterion or use equal weighting.

Step 5. Determine how the individual alternatives will be ranked.

- *Forced choice*: Each alternative is ranked in relation to the others. The alternative best meeting the criterion gets a score equal to the number of alternatives and the worst would get a 1.
- *Rating scale*: Each alternative is rated independently against an objective standard. For example, a 1–10 scale would have 1 = very

low (does not meet the standard at all) and 10 = perfect (absolutely meets the standard).

- *Objective data*: Enter actual data, rather than the opinions of the individual(s) doing the ranking.
- *Yes/No*: If the criteria are expressed in absolute terms, so an alternative either meets the criterion or not, simply enter Y or N to indicate conformance or nonconformance.

Step 6. Review the results and take action as required.

Guidelines and Tips

Whenever comparing alternatives (forced choice method or rating scale), the group must agree on the relative importance of the alternatives/criteria for scoring purposes. Relative importance can be established either through consensus discussion or through voting techniques. You will usually want to reach agreement rather quickly on this. The amount of time you spend should be based on the importance of the problem/solution and on the number of alternatives and criteria. If there are a large number of alternatives/criteria, you can reach agreement quicker keeping in mind the impact of each individual item on the list is smaller.

Depending on the nature and impact of the problem, this process can be simplified for quicker and easier use. For example, the process can be simplified by assuming that the criteria are of equal importance and therefore the ranking of alternatives can be skipped. You can look for other simplifying assumptions. Just be aware of their impact on results.

There is no one best way to weight criteria or alternatives. In the forced choice method, you rate each element against the other, based on the number of choices. This is a time consuming method, though. The rating scale method is quick, but has the drawback that people tend to rank every criterion as very important or high on the scale of 1 through 10.

Again, there is no one best method. Use the method that provides you with the most information. Before using any of the alternative approaches described here, however, think about the implications of the various schemes. If you plan to use prioritization matrices repeatedly, you might set up a simple spreadsheet to assist you with some of the calculations.

EXAMPLES

The examples are included in the text.

SOFTWARE

Some commercial software available includes but is not limited to

- Edraw max: http://www.edrawsoft.com
- Smartdraw: http://www.smartdraw.com
- Affinity Diagram 2.1: http://mobile.brothersoft.com/
- QI macros: http://www.qimacros.com

REFERENCE

Harrington, H.J. and Lomax, K. *Performance Improvement Methods.* New York: McGraw-Hill, 2000.

SUGGESTED ADDITIONAL READING

Asaka, T. and Ozeki, K., eds. *Handbook of Quality Tools: The Japanese Approach.* Portland, OR: Productivity Press, 1998.
Brassard, M. *The Memory Jogger Plus.* Milwaukee, WI: ASQ Quality Press, 1989.
Eiga, T., Futami, R., Miyawama, H., and Nayatani, Y. *The Seven New QC Tools: Practical Applications for Managers.* New York: Quality Resources, 1994.
King, B. *The Seven Management Tools.* Methuen, MA: Goal/QPC, 1989.
Mizuno, S., ed. *Management for Quality Improvement: The 7 New QC Tools.* Portland, OR: Productivity Press, 1988.

19

Mind Mapping/Spider Diagrams

Frank Voehl

CONTENTS

DEFINITION

Mind mapping is an innovation tool and method that starts with a main idea or goal in the middle, and then flows or diagrams ideas out from this one main subject. By using mind maps, you can quickly identify and understand the structure of a subject. You can see the way that pieces of information fit together in a format that your mind finds easy to recall and quick to review. They are also called *spider diagrams*.

USER

This tool is best used by individuals, but can also be used with a group of almost any size up to 10.

OFTEN USED IN THE FOLLOWING PHASES OF THE INNOVATIVE PROCESS

The following are the seven phases of the innovative cycle. An X after the phase name indicates that the tool/methodology is used during that specific phase.

- Creation phase X
- Value proposition phase
- Resourcing phase
- Documentation phase
- Production phase X
- Sales/delivery phase X
- Performance analysis phase

TOOL ACTIVITY BY PHASE

- Creative phase—Mind maps and spider diagrams are often used during the creative phase to analyze a problem and create a list of potential solutions. They are also used to pictorially show the different aspects of the situation to be sure that all of them are addressed when the solution is prepared.
- Production phase—Mind maps and spider diagrams are used to visibly depict the different aspects of a problem to be sure that all of them are addressed.
- Sales/delivery phase—Mind maps and spider diagrams are frequently used to visibly depict all approaches to marketing and sales campaigns, and the advantages and disadvantages of each.

HOW TO USE THE TOOL

During the late 1960s, Tony Buzan developed his studies of brain function into the mind mapping technique as we know it today. Mind mapping combines the visual and spatial aspects that enable the brain to be most efficient, giving new clarity, greater control of your thinking processes, and therefore much greater control of your life. Hundreds of millions of people worldwide have adopted mind mapping as an efficient, stimulating way to organize thoughts, capture beliefs, generate ideas, and plan almost anything.

Buzan teamed with Chris Griffith to build the Open Genius concept to innovation and creativity. The mind mapping approach is built around four principles for applying innovative thinking*:

- Extending the mind mapping philosophy and approach globally for the benefit of humankind
- Becoming a recognized expert in maximizing creative potential and generating innovative solutions *that work*
- Learning a practical, repeatable process for innovation that can be implemented by anyone, from individuals to multinational organizations
- Gaining the knowledge and skills to facilitate powerful applied innovation strategies within an organization or for clients
  - Certifying others as *OpenGenius practitioners* in applied innovation

Applied creativity and innovation skills are no longer nice to have, says Buzan—they are a *must-have* and are needed to distinguish innovators in

---

* Innovative thinking can benefit greatly from mind mapping because it is able to consume all the common skills found in imagination, creativity, flexibility, and the organization of your ideas. There are some fundamental principles that you need to know if you want to improve your creative thinking. According to psychological research, these include using shapes, colors, unusual elements, and dimensions. Although people are gifted with innovative thinking at birth, they lose this ability and these people can take advantage of mind mapping because it suits them perfectly, since mind maps consume skills found in imagination, creativity, flexibility, and organization of ideas. According to many psychological researches, creative thinking has elements that include the use of shapes, colors, unusual elements, and dimensions. Aside from that, creative thinking can also adjust conceptual positions and create response to appealing objects.

an increasingly competitive market by learning the recipes for innovation on demand, as the history of mind mapping illustrates.*

Figure 19.1 is a hand-drawn mind map by Paul Foreman used to show the application of mind mapping and innovation.†

Figure 19.1 is used for planning or creating and advancing thought-provoking ideas centered on the theme of innovation. Spider diagrams are similar to mind maps as they are extremely organized, and use thought clusters and webs in a neat and clearly structured layout approach. As with mind maps, spider diagrams start with a central idea and branch out from there. The main differences are that spider diagrams do not always use color, nor is there any particular way of how you should structure a spider diagram, as shown in Figure 19.2.

Figure 19.2 shows a generic spider diagram, and emphasizes the advantages of spider maps, which is that they are extremely simple in concept, making them very easy to make. You can draw them in any manner you like, and they will be done in a matter of minutes. Spider diagrams are not the same thing as mind mapping, but they can be just as useful in many ways, to use as an idea organizer or an alternative, and they are a very visual note-taking method.

The reasons why you should use spider maps over linear notes is that, as with mind maps, they reflect on the structure of the brain, and start with a central idea, branching outward from the center like a lotus blossom, creating many branches and endless possibilities. Before mind maps were invented by Tony Buzan, simple spider diagrams would have been the closest thing to them. Sometimes, it can be a bit of a job to create a whole

---

* Although historians continued to find various traces of mind mapping throughout history after Leonardo da Vinci first used a form of the method, they were relatively insignificant until the beginning of the 1950s or 1960s. Network semantics was developed in the late 1950s, just a basic theory in order to understand the demonstrations of how human beings develop learning. This concept was furthered by Dr. Collins in the 1960s. Dr. Collins was considered as the father of modern mapping because of his extensive commitment to publish research papers about his creativity, graphical thinking, and learning. Ross Quilian and Alan Collins were the ones who shaped the future of mind mapping. They both used a kind of network where all the concepts and ideas were related by links that would show them how a certain object is related with another. This is how mind mapping became very useful in learning, sharing concepts, and other various collaborative techniques. But it was not until the 1960s that a British psychologist, Tony Buzan, made its use very popular, and thus became known as the *father* of modern mind mapping; he even created a set of rules to be used when applying the mind mapping method and its techniques.
† See www.mindmapinspiration.com. If you use any free mind map downloads on your blog or website, please link back to the Mind Map Inspiration website at http://www.mindmapinspiration .co.uk or the Mind Map Inspiration blog at http://www.mindmapinspiration.com.

FIGURE 19.1
Mind map or spider diagram.

mind map with all the color and peripherals. If you just want to brainstorm some ideas, this is where you can just as easily use spider maps.*

Putting the Tool to Use

Mind maps can be produced in two methods or ways, the first being a manual technique where the maps are hand drawn, and the other being drawn with the use of software. With manual techniques, there is a learning curve that must be overcome to generate maps effectively, as it takes

---

* Spider diagram mapping (which is sometimes called *semantic mapping*) is a graphic organizer that can be used for brainstorming ideas, aspects, and thoughts of normally a single theme or topic. It gets its name because of the way it looks when drawn out. It is typically done for writing stories, papers, and research brainstorming.

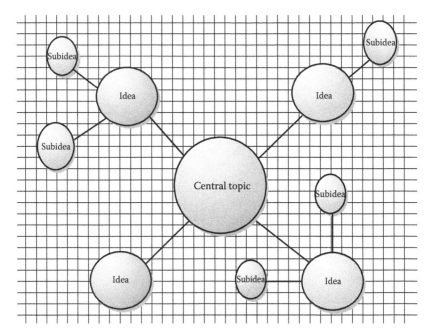

FIGURE 19.2
Generic spider diagram.

some practice in defining locations and placement of topics and details. It has been noted that the use of colors and symbols can positively influence the information retention of the presented information. If the maps are to be shared, photocopying or scanning must be used, which can slow the transfer of the map to others. It also makes the maps a reference document, not a living document for changes.

Mind mapping has seen a growth in interest and use based on the ability to create the maps in an electronic digital format; this is the second way in which maps can be generated. The electronic software that is now available allows the maps to be created quickly and reliably. These electronic maps offer the same value as manual maps with the added benefits of being easily exchanged or viewed during courses or meetings, while also allowing for the inclusion of other artifacts such as documents, drawings, pictures, multimedia, and Internet links. Teams can use one map as a common placeholder of information and use it to grow the map into further detail, solutions, or ideas. This makes it a natural tool for creativity and innovation. These features have strongly influenced the acceptance of mind mapping not only on an individual level but also as a worthwhile tool in businesses.

Tony Buzan, the creator of mind maps, documents how they are a powerful creative approach to note-taking. Mind maps are known by various names, such as mind mapping, concept mapping,* spray diagrams, and spider diagrams, to name a few. Innovation and creativity go hand in hand, and you cannot push forward and create a space for innovation within your company without first finding a way to enhance your collective creative skills, states Buzan—"Wherever creativity goes—and, by extension, wherever talent goes—innovation and economic growth are sure to follow."

If you need to brainstorm an issue, or design and develop a new product, or for the *soft sciences* deliver a marketing campaign, or make a pitch for a business concept or deal, or simply use a visual method to discuss a problem, creativity is a key, and for many innovators, creating a mind map is an excellent way of coming up with creative ways to address these matters. Buzan outlines a simple five-step mind mapping process for you to follow to create your own brainstorming mind maps and boost innovation in your organization or company.†

Five-Step Mind Mapping Process

Step 1: Start by Drawing a Quick-and-Easy Mind Map Burst

Begin your mind map by outlining in very basic terms your central idea, theme, or topic. Choose a stimulating central image that represents the theme or topic you are addressing. From your central idea, begin to capture every idea that comes into your mind on that subject, dividing them into key ideas, or main brain branches. From

---

* Concept maps are useful tools in taking notes and representing them in a graph. Concept networks are also referred to as *knowledge maps*. The different networks usually consist of links and nodes. Links are the ones representing the relationship between different concepts and nodes are the ones representing the different concepts. Concepts are sometimes linked with labels. Links can be in forms of one-directional, bidirectional, or nondirectional. Generally, links and concepts can be categorized in such a way that they can be specified, associated, and divided into categories like temporal or even causal relations.

† Tony Buzan is the world-renowned popularizer of mind mapping and expert on the brain, memory, speed reading, creativity, and innovation. He has been named as one of the world's top 5 speakers by *Forbes* magazine. Through more than 40 years of research into the workings of the brain, Buzan has dedicated his life to developing and refining techniques to help individuals think better and more creatively, and reach their full potential. In the course of his work, he has awakened the brains of millions worldwide. Described as "one of the most influential leaders in the field of thinking creatively," he utilizes his accredited training courses to build a network of highly specialized experts in creative thinking, memory, and speed reading techniques. Source for more details: http://thinkbuzan.com/articles/view/mind-mapping-for-innovation/.

your main ideas and branches, add your subideas, or child branches, by letting the ideas flow as fast as possible. Having to work at this type of speed unchains your brain from habitual thinking patterns, and encourages new ideas and innovation. Many of these new ideas may at first seem absurd; however, they often hold the key to new perspectives, and offer many clues for breaking old habits. Buzan feels that the most innovative solutions come from the kernel of an idea, and you want to encourage as many new thoughts and creative ideas as you can at this stage.*

Step 2: Do a First Reconstruction and Revision

Next, take a short break or move onto another task, allowing your brain to rest and begin to integrate the ideas generated thus far. Then create a new map in which you identify the core ideas/branches, categorizing, building up hierarchies and finding new associations between your preliminary ideas. Similar ideas may be repeated in different areas of your map, on different branches. These peripheral repetitions reflect the significance of the repeated idea as it is influencing multiple aspects of your thinking and should be pursued further as a lead to innovation.

Step 3: Incubation

After completing the above steps, take a break and let it sink in.†
Most innovation comes from those *ah-ha* moments, and these sudden creative realizations often occur when the brain is relaxed such as daydreaming, sleeping, or running. This is because such states of mind allow what Buzan calls the "the radiant thinking process to spread to the farthest reaches of the brain, increasing the probability of mental breakthroughs." Using relationships,‡ mind mapping enhances the potential of focused daydreaming to help facilitate innovation, and should not be underestimated, as Albert Einstein

---

* You can learn more about these basics of mind mapping with Tony Buzan's *Create a Mind Map* guide: https://www.youtube.com/watch?v=tAUsZ9eiorY.
† Linear thinking is the concept wherein these incubation elements are present. In general mind mapping, analyzing is only secondary. What is important is the very thing that pops into your head while focusing on the idea. But you have to be sure that whatever pops into your mind is directly related to the topic, otherwise you have to use your reasoning ability to pattern your thoughts; thus, the use of incubation to let things sink in.
‡ Like familiar strangers with whom you are connected because of a friend of another friend, that is how facts and information can be accurately intertwined with one another in a mind map. You can use arrows, colors, lines, and even braches to be able to show the possible connections of ideas being generated inside your mind. Relationships of one idea from the idea are important for one to understand the derivative of the new information patterned out through construction of the map.

did it himself to help visualize and arrive at his famous theories in their primitive raw forms.*

Step 4: Do a Second Reconstruction and Revision

After the initial incubation, your brain will have a fresh perception on your first and second mind maps, so Buzan feels that it will be useful to do another what he calls *quick-fire mind map bursts.* During this reconstruction stage, you will need to consider all the information gathered and integrated in stages 1, 2, and 3 of the process for generating innovation, in order to make a comprehensive final mind map.

Step 5: Complete the Final Stage

Using your final mind map, you now need to search for the solution, decision, or visualization, which was your original creative goal. This often involves doing further work on your mind maps by making connections between branches, and also perhaps adding further subbranches where needed. When using iMindMap (see section on Software), you can move branches around your workspace, which often stimulates more connections and ideas. The answer may not present itself straight away, so be patient; take another incubation break if you need to. Making additional notes and finding connections and patterns that you had not previously been aware of can lead to major new insights, breakthroughs, and genuine innovations in your thinking.†

Typical Example of Mind Maps

The following is an example of some of the main features of a mind map to create a better Earth:

---

* Creative individuals are considered fortunate because not everyone can bring out their creativity especially as they grow older. Being creative is natural among children because they are using to the fullest their brain's right hemisphere, but as a person grows older, they become more rational and structured, and find themselves using mostly their left brain hemisphere. Artists, for example, are gifted people, and so they find it easy to visualize tasks or ideas without great effort.
† Some of the most significant mind mapping notes are those that have spaces left for further note-taking add-ups. You can also choose to highlight some of the important information that the subject at hand demands. Here, proper spacing is good to provide additional information to join into the picture. A good mind mapping is one that is overflowing with appropriate details from the past and present.

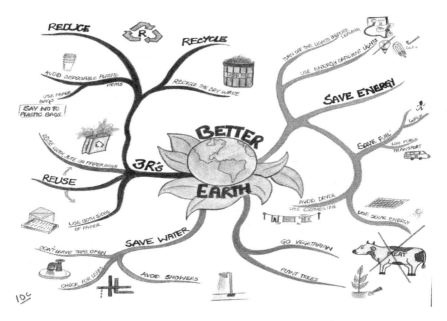

FIGURE 19.3
Mind map for a better Earth.*

1. Notice the title of the subject or image being explored in the center of the page, and the circle around it. This is shown in Figure 19.3 for how to create a better Earth.* Next, draw the main branches, such as Save Water, 3 Rs, Save Energy, and Plant Trees, etc.
2. The person or team uses single words or simple phrases. Many words in normal everyday writing are nothing more than a form of padding, and are used to ensure that facts are conveyed in the correct context, and in a format that is pleasant and, hopefully, easy to read. In mind maps, single strong words and short, meaningful phrases can convey the same meaning more potently. Excess words just clutter the mind map.
3. Printed words are used, as joined up or indistinct writing is more difficult to read.
4. The example uses color to separate different ideas, along with symbols and images. This helps separate ideas where necessary. It also

---

* Description: Kartik Agarwal, Springdale Senior School, Punjab. This is one of the top entries in the mind mapping competition run by the City Montessori School, in Lucknow, India. Topics for the mind maps included (a) innovative ideas on inspiring global understanding, (b) innovative ideas for holistic wellness, (c) innovative ideas for divine education, and (d) innovative ideas to create a better Earth. Eighty mind maps were received from around India, and we are proud to be displaying 14 of these on the Mindwerx Resources page at http://www.mindwerx.com/mex/mind -map/hand-drawn/6640/kartik-better-earth.

helps the creators to visualize the mind map for recall. Color can also help demonstrate a certain organization of the subject. Symbols and pictures can help you remember information more effectively than words, so, where a symbol or picture means something to you, use it.*

Advantages

- Mind mapping and diagramming act as trigger devices for your brain, creating an explosion in creativity, innovation, and knowledge sharing.
- These diagrams are based on the fundamental principles of creativity, and so they are ideally suited to supporting ideas generation and innovation.
- They tap into your rational and grounded skills and your imaginative and free-associative skills, encouraging them to work together to both amplify and focus your thinking in a divergent and convergent manner, creating the ideal setting and space for innovation.
- They allow you to view a great many elements all at once, thus increasing the probability of creative association and integration, and consequently, innovation.
- Mind mapping allows you to assess risks and represent a tremendous amount of information in a relatively small space. You can have all your notes for a topic on one piece of paper, with your ideas arranged in a way that encourages you to see relationships between them.†
- Innovation is about originality and achieving a cutting edge. Mind maps encourage your brain to track out ideas that normally lie in relative obscurity on the edge of your consciousness and thinking.

---

* Leonardo da Vinci is considered by many to be one of the world's most famous artists, and studies reveal that he was ambidextrous. Some say that da Vinci utilized mind maps to complete his works of art. Because of this, a lot of artists as well as writers are now considering the use of mind maps in completing their works. Focus is a very important factor in completing art projects, and mind mapping can aid artists in keeping themselves focused despite interruptions or distractions.

† When you want to create a mind map to assess risks, it is often a very simple process. Start with your idea in the middle of a paper, and then start drawing out the risks—personal, financial, or operational risks. You will notice that you will be able to generate more ideas that way. Then, for every risk you listed, try to provide solutions for each of them. Repeat the process till you run out of idea. Using this approach, after you complete the mind map, you will be able to figure out whether the idea was risky to begin with. If you can provide solutions to all (or a majority) of the issues, then you can understand that the risks are often manageable. On the other hand, if the amount of risks exceeds that of your solutions, then you are better off not trying it without further imagination and ideation.

- Diagramming increases the probability of you straying a bit far from the norm, exploring unconventional routes and producing truly innovative ideas and concepts.
- Generate more ideas as mind mapping allows you to start quickly and generate more volume in less time.
- You do not have to edit or order your thoughts; just start with a creative center icon or doodle in the middle of your page and begin printing words on lines emanating from the center as you think of them.
- The diagram uses branching with a free-flow format, adding words to one branch one moment, then skipping over to another branch the next, which seems to increase your chances of generating new ideas.
- Improve your memory as devices for helping one remember material becomes much easier. Colors, images, and key words, three central ingredients of mind maps, are much more engaging to the brain than sentences. A well-made mind map is almost impossible to forget.*
- Use your whole brain. Mind mapping helps you strengthen your analytical left brain by training you to look for the most essential key words. At the same time, it stimulates the right brain by encouraging you to use colors and images.

Disadvantages

- Mind mapping can sometimes become a very time-consuming process, especially for beginners who are just starting to use this tool, and may need a seasoned facilitator.
- Those people who are very logical and do not tend to trust their intuitive side will oftentimes find it difficult to intuitively and easily use this tool/technique, because their logic is telling them that this is a different way of working, and is impossible. On the positive side, the habit of utilizing the brain's multiprocessing capabilities in this way starts to develop after the first half-a-dozen or so attempts, and so almost everybody can use it with a little perseverance.

---

* Sometimes, designers of mind maps get so carried away that the end result does not make sense any more. Maps that are too cluttered are quite difficult to fix and, worse, the person can get stuck on an idea and has no option of going forward or moving back. Generally, your ideas must be briefly keyed inside your map. Your topics should be summarized statements that are persuasive enough like news headlines. If you ever want to add details and additional information under that category, leave spaces so that you can include it when that time comes.

- Mind maps provide a structured way for recording information and therefore they are not suitable for recording more chaotic forms of creative idea production, that is, ideas generated during brainstorming sessions. In this case, mind maps would slow things down and may inhibit idea production through the effort of the participants to associate and structure their ideas on paper. On the positive side, mind maps could be beneficially used as a follow-up stage in a brainstorming session for sorting out and evaluating the ideas that have been recorded, much as with a KJ (affinity) diagram.
- Mind maps provide a self-explanatory presentation of information and structure only to the creator and those participants who took part in the initial development of the mind map, in that meaning and symbol use were negotiated within the mind mapping group and participants developed a shared understanding. In other words, certain words, associations, and structures that represented a meaning for their creators do not necessarily mean anything to someone who did not sit in on or take part in the diagram's creation.

SUMMARY

Mind mapping and spider diagrams are unique ways of organizing your thoughts, and they let your brain naturally branch out into subcategories and sub-subcategories, while keeping the idea flowing and growing. The mind map is about as brain friendly and organic as it gets. It works with your brain, and for your brain, and is an extremely effective method of creating a flow of ideas and of taking notes. Not only do mind maps show facts, they also show the overall structure of a subject and the relative relationship of individual parts of it. They help you to associate ideas, think creatively, and make connections that you might not otherwise make. Mind maps are useful for summarizing information, for consolidating large chunks of information, for making connections, and for creative problem solving.

First you write what the topic/concept of the brainstorm is in the middle of the page, then draw a bubble around it. Next, start with your subidea. You add the subidea by drawing an arrow or a line from the bubble outward. You then label it at the end of the line. Some people like to make the line an arrow, and others prefer to draw a circle around the label. There are no set *stick to these or else* rules in mind maps and spider diagrams. Mix and match, and trust your instincts to do it your own way. Add subheadings, sub-subheadings, and so on. The ideas should start to flow and *grow.*

EXAMPLES

Examples included in the section of this chapter entitled "How to Use the Tool."

SOFTWARE

A list of 24 essential mind mapping and brainstorming tools is found at the website http://mashable.com/2013/09/25/mind-mapping-tools/.

The following is a description of five of the most popular software packages.

- iMindMap

    You can stimulate innovation in half the time with mind mapping software. Tony Buzan's iMindMap utilizes the true principles of mind mapping and makes creating maps refreshingly simple. Editing and revising maps takes no time at all, so there is no need to start from scratch when you need to rearrange or expand your map—cutting down significantly on the steps above. Speed Mind Map Mode draws the map for you so you can capture ideas as they come without interrupting your train of thought. iMindMap will unleash your creativity and enrich your business with innovation and success.

- Mindjet MindManager

    With MindManager, brainstorming is just the beginning. It is a powerful, productivity-boosting, project-building mind mapping software. Capture ideas. Manage meetings. Create strategic plans. Organize anything. MindManager has the robust features you need to get everything you do done better. See more at http://www.mind jet.com/sem/mindmanager/?_bt=68750195231&_bk=mind%20map ping%20software&_bm=e&gclid=CjwKEAjw3_ypBRCwoKqKw5P

9wgsSJAAbi2K9_8126tEnvHsOspeCpDxPhjPPPVnpte5yFNJIoUwR
VhoCqWnw_wcB.

- MindGenius

    MindGenius is business mind mapping software that helps you capture, visualize, and manage your ideas and information. The visual layout of information enables projects and tasks to be scoped and agreed effectively. Teams will make better, more informed decisions and take action quicker. See more at http://www.mindgenius .com/benefits#sthash.2sORaSXV.dpu.

- MindMeister

    MindMeister is considered the best online mind mapping app currently on the market. With its award-winning online version and its free mobile apps for iPhone, iPad, and Android, users are able to mind map at school, at home, at the office and even on the go. Mind mapping with MindMeister is so simple and intuitive that anyone from first graders to chief executive officers can use it to improve their productivity and turn their creative ideas into action. MindMeister offers a number of powerful features that let users collaborate and brainstorm online, plan projects, develop business strategies, create great presentations, and utilize the enormous power of mind maps for their education. See more at https://www .mindmeister.com.

- XMind

    XMind has been around for a good long time, but it has not lost its power; it is still extremely flexible, works great on any desktop OS, and makes it easy to organize your ideas and thoughts in a variety of different styles, diagrams, and designs. You can use simple mind maps if you choose, or *fishbone*-style flowcharts if you prefer. You can even add images and icons to differentiate parts of a project or specific ideas, add links and multimedia to each item, and more.

SUGGESTED ADDITIONAL READING

Bullard, R. Engage the brain. *The Times Educational Supplement*, Teacher Supplement, pp. 14–15, June 23, 2006.

Hay, D. Making learning visible. *The Role of Concept Mapping in Higher Education*, vol. 3, pp. 295–311, 2008.

Lamb, A. Digital dog ate my notes. *Tools and Strategies for 21st Century Research Projects*, no. 2, pp. 77–81, 2009.

Makany, T. Optimising the use of note-taking as an external cognitive aid for increasing learning. *British Journal of Educational Technology*, vol. 4, pp. 619–635, 2009.

Nast, J. *Idea Mapping: How to Access Your Hidden Brain Power, Learn Faster, Remember More, and Achieve Success in Business.* Hoboken, NJ: John Wiley & Sons Inc., 2006.

Petro Jr., N. Hate taking notes? Try mind mapping. *GPSolo*, vol. 4, p. 23, 2010.

Rana, E. Let your chart rule your head. *People Management*, vol. 8, no. 3, pp. 42–43, 2002.

Tergan, S.-O. Mapping and managing knowledge and information in resource-based learning. *Innovations in Education & Teaching International*, vol. 4, pp. 327–336, 2006.

Thane, P. It's a steal. *The Times Educational Supplement*, pp. 32–33, June 6, 2008.

20

Online Innovation Platform

Lisa Friedman

CONTENTS

Expand the possible.

DEFINITION

Online innovation management platforms enable large groups of people to innovate together—across geographies and time zones. Users can post ideas and value propositions online, and can collaborate with others to make these stronger. The community can rate and rank ideas or value propositions, post comments and recommendations, link to resources, build on each other's ideas, and support each other to improve each other's innovations.

In addition to the broadly visible *front end* of platforms designed for innovators, most online innovation management platforms also provide a *backroom* or *innovation dashboard* where leaders can track and analyze the innovations occurring in the larger group. A typical dashboard shows leaders how many ideas or value propositions are created over time, which departments or groups contribute, which groups comment and make innovations stronger, how many ideas turn into pilot projects, and how many become commercially viable products or services in the marketplace. Many platforms enable leaders to cluster and analyze the overall portfolio of projects, to see how many have short-term versus long-term results, how many are high risk versus low risk, the number of innovations in each phase of implementation, the partners or suppliers involved, or the overall projected value of their innovation portfolio.

USER

This tool is designed to enable groups of innovators to collaborate online. A group can be virtually any size, from 5 to 50 users up to many thousands.

Two major groups benefit from using an online innovation management platform:

- *Innovators* can ensure their ideas are visible to others. They can collaborate across teams or departments or even company boundaries. They can tap into the intelligence from a larger group to enhance their own innovations as well as supporting others. The result is an active innovation community with a culture of continuous idea exchange and excitement about creating a positive future.
- *Innovation leaders* can use the tool to help create this engaged innovation community within their organization or network. They can invite innovators to create solutions in the areas that are most strategically important to their enterprise. They can also track and analyze their overall innovation pipeline or portfolio.

OFTEN USED IN THE FOLLOWING PHASES OF THE INNOVATION PROCESS

- Creation phase X
- Value proposition phase X
- Financing phase X
- Documentation phase X
- Production phase X
- Sales/delivery phase X
- Performance analysis phase X

TOOL ACTIVITY BY PHASE

Online innovation management platforms are most often used in the first three phases of the innovation process: the creation phase, value proposition phase, and financing phase. In these phases, innovators learn about a challenge at hand; search for ideas created to date; generate their own ideas; turn ideas into value propositions; collaborate with others to enhance their value propositions; iterate and optimize their value propositions, even to the extent of *pivoting* to entirely new ways of achieving the intended goal; and connect with those who can approve, fund, and support their innovations.

While using online innovation platforms in the first three phases is much more common, the platforms can also be used in each of the remaining phases as well. The platform enables users working in any phase of innovation to post their work and to collaborate with a larger community to get input, ideas, contacts, and resources to improve their innovations.

HOW TO USE THE TOOL

Online innovation management platforms can be used to efficiently tap into group intelligence within or outside an enterprise to enhance any type of innovation, including:

- The next generation of new products and services
- Improvements to existing products and services
- New business models
- New branding concepts
- New customer experiences
- New operational efficiencies and process improvements
- New ways of engaging employees
- New safety practices
- New ways to create green or sustainable solutions, or to enhance the environmental or social responsibility of existing products and services
- Solutions to problems that need to be solved or anything new that needs to be created or improved

Spontaneous versus Sponsored Innovation

Spontaneous innovation. Online innovation management platforms can be used for either spontaneous or sponsored innovations. In spontaneous innovation programs, users throughout an organization or innovation ecosystem can generate ideas on their own and post these to the platform. (This is sometimes referred to as *bottom–up* innovation, where people working on the frontlines generate ideas from their specific knowledge and expertise.) Users post what they believe ought to exist or what they most want to create. By using an innovation platform that engages a large community, their ideas become visible to others. Others can rate and rank their ideas, or comment and build on them. Ideas that generate the strongest group interest and support rise to the top.

Sponsored innovation—innovation challenges and campaigns. In an era that is becoming increasingly global, social, mobile, and more green/sustainable, many industries are being disrupted. Existing products and services may need to be produced or offered in entirely new ways, while the race is on to create the next generation of products and services. In this rapidly changing business environment, leaders can source strategic new solutions by launching innovation campaigns or competitions in the specific areas where innovation is needed most. The platform then enables employees across a wide range of roles and job descriptions to contribute their ideas and to build on the ideas of others. Campaigns or innovation challenges can also be opened up to partners, suppliers, or customers, and there could be open campaigns to engage the public at large.

To launch an innovation campaign, a leader may issue an online *Invitation to Innovate* through text or video on the platform. Many platforms enable leaders to post videos or articles to help potential innovators understand the challenge at hand. Information might cover emerging trends in the marketplace, important customer needs, new technical information, or solutions being generated by others. Facts and statistics can be helpful, or videos of customers talking about what frustrates them and what they want most, to bring the challenge alive.

Examples of Campaign Questions

- A book publisher might ask, "How can we make books more interactive in order to engage teenagers who are used to an interactive mobile digital world?"
- Education leaders could ask, "How could we engage community members to enhance our students' learning?"
- Hospital leaders might ask their staff, "How can we reduce infections?" Or they might ask a wider network of patients, families, and technology leaders, "Now that more of our patients have smartphones or other personal devices that track their exercise, heart rate, sleep patterns, blood pressure, blood sugar, etc., how can we integrate this personal data into their medical care?"

Online innovation campaigns are generally time limited. They may run for three weeks to a month, and sometimes even less (so-called sprints). There are often multiple rounds of feedback within a campaign, where innovators post ideas and get feedback in the first round of the campaign, and then post their revised solutions in the next. The final competition for winning ideas may come after one or more rounds of revision, to enable innovators to integrate what they learn from the wider community. Ideas are rarely at their best from the beginning, and an important benefit of the software is that it enables innovators to use the wisdom of the crowd to improve their ideas over time.

Some organizations will assign individual leaders to sponsor separate sequential campaigns, so there is always an online strategic campaign going on. A campaign gains energy when there is a committed innovation sponsor who plans to put the winning ideas to work. As an example, a company that develops retirement communities for seniors might ask,

- "How can we use technology in our residences to help seniors *age in place*, to live independently in their own homes?"
- Following this campaign, another leader might sponsor a new campaign that asks, "How could we design our interiors using healthy materials?"
- Later, another leader may initiate a campaign that asks, "How can we design our community to enhance residents' connection with family and friends?"

Turning Ideas into Value Propositions

Some platforms also educate users to develop their ideas into value propositions, where simple ideas gain a smaller number of points and value propositions, which require more effort to create and are much more developed, win significantly more points. Several platforms offer a standard template for creating value propositions, to create a shared innovation language and practices across the organization. Widely shared methods enable innovators throughout an organization to collaborate more quickly and effectively. Innovators and teams fill in the template online, and other users can then comment on specific categories within the template. For example, customized versions of both Qmarkets and IdeaScale innovation platforms offer CO-STAR value proposition templates. These templates guide innovators to provide the detailed thinking to turn their initial ideas into value propositions, by specifying

- *C—Customers.* Who are the intended customers and what are their interests or most important needs?
- *O—Opportunity.* Given emerging trends, what is now a significant, high-impact, market opportunity?
- *S—Solution.* What are the details of your solution? A fresh CO-STAR may give a high-level conceptual design, while later, more developed CO-STARs would give more details about the solution (e.g., how it would be developed, the business model, a customer acquisition strategy, a go-to-market plan, or a detailed financial plan).
- *T—Team.* Who are the team members and what is their expertise?
- *A—Advantage.* What are competing solutions and what are the advantages of the proposed solution over these alternatives?

- *R—Results.* What results will be achieved for key stakeholders throughout the innovation ecosystem? What are rewards for customers? What are the results and return on investment for the team, company, or investors? Would there be rewards for key partners or suppliers, or for the community, the environment, or society?

Useful Features of Online Innovation Management Platforms

Platforms that supply such templates lead innovators to develop their ideas into more in-depth value propositions and enable users who comment to give more specific and useful input. For example, some users may know team members who could be a great help to the project, or others may know customer groups who would like to run pilot projects, etc.

Gamification and Rewards

Many online innovation platforms use gamification techniques and rewards to increase engagement—both for innovators posting ideas and value propositions, and for the wider community who comment and help make innovations stronger. Typical methods include

- *A point system.* Users gain points by submitting ideas and value propositions, as well as by commenting or collaborating with others. The most highly rated innovations or the most active commenters rise to the top and become visible to the whole community.
- *Virtual currency.* Some organizations design their innovation platforms to turn points into virtual currency that can be redeemed for real rewards in an online store.
- *Nonmaterial rewards.* Rewards may be nonmaterial as well, in that some innovators prefer time with an executive or an expert in the field of the innovation they are working on—time to be heard, to ask questions, to pitch their idea to a group that could approve or support it, or time that might lead to a longer-term collaborative working relationship.
- *Implementation.* One of the most valuable rewards for most innovators is seeing their ideas become reality, to receive the recognition that goes along with this, and to gain the opportunity to continue

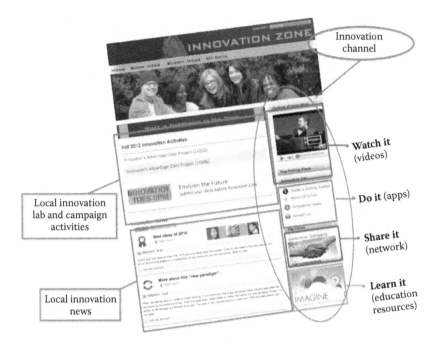

FIGURE 20.1

Features of online innovation management platforms.

working on their project or to receive genuine interest or support for their next new idea. Figure 20.1 points out some of the useful features of online innovative management platforms.

Software is continually being upgraded, so this list of typical features will undoubtedly change over time. Typical features currently include

- User profiles/knowledge management. Users can post their interests and expertise, including areas that may not be evident from their current job descriptions.
- Insights and information sections. Particularly related to specific innovation challenges, but there also can be links to industry news or innovation in other fields that may spark ideas for innovators.
- Idea search. Search for similar ideas and clustering similar ideas together.
  - This can occur through individuals tagging their entries with keywords.

- Several platforms do this as well through automated or *semantic search*, where the software identifies similar language and concepts and then groups these together.
- A few platforms generate mind maps to help illustrate how ideas are connected.
- Posting questions (and answers)
- Posting ideas for innovations
- Posting value propositions
- Comments. These might include
  - Ways to make the idea itself stronger
  - Ways to strengthen how the idea is communicated
  - Links to helpful information or resources
  - Suggestions for additional team members or advisors
  - Connections to funding or other support
- Rating and ranking. Most platforms offer several ways to evaluate ideas and value propositions and they recommend using a variety of methods:
  - A simple thumbs up or *Like* approach for favored ideas
  - A more detailed number of stars, or a 1–5 or 1–10 rating
  - A scorecard with stars or number rating on multiple criteria
  - Token voting, where users get a set number of tokens and can allocate these among ideas
  - Idea tournament or *pair-wise voting*, where users are given two ideas and they choose their favorite. This one-to-one comparison continues until all the ideas have been matched and the winners are put against other winners (as in a sports tournament), until one final winner emerges.
  - Prediction markets or idea trading. Users predict how well they think ideas will do over time, and trade ideas with others in a way that assigns more value to the ideas they think will do best over time.
- Evaluation and decision making. Value propositions that rise to the top of the lists through crowd voting will often go to innovation leadership teams, managers, or teams of experts for further evaluation and decision making.
- Project management tools. Several platforms track winning value propositions through concept, design, prototyping, pilot, business planning, launch, scale-up, and implementation phases.

- Reporting and analytics. Many platforms synthesize real-time information and trends over time for multiple areas of innovation, including
  - Numbers of innovations proposed by specific groups, such as departments, level of employees, locations, etc.
  - Participation over time
  - Summary reports, such as all comments for one particular innovation, all innovations proposed by a given department, or all innovations proposed in a specific campaign
  - Return on investment
  - Innovation pipeline or portfolio summaries
    - How many projects are in each phase, such as concept, prototype, pilot, launch, etc.
    - Risk levels: how many projects are high risk, medium risk, low risk
    - When projects are expected to go to market: three to six months, one year, two years, or longer-term projects
    - How many innovations will be offered to specific customer groups or market segments
    - Financial reporting: overall project valuations, as well as projected costs and returns over time

Important Requirements for Using Online Innovation Management Platforms

Software alone cannot create a dynamic enterprise with strong innovation capabilities. An innovation platform will only be useful when paired with strong leadership that clarifies a guiding innovation strategy for the enterprise. Innovators must know where they are headed to create the value needed most. Leaders also need to be actively engaged in using the platform, as the wider community will model their level of participation.

The platform will also be most effective when there is a clear innovation architecture and organization design, when innovators understand

- How to submit ideas and who selects winning ideas (The crowd? Their manager? An expert panel? Customers? Investors?)
- How long a campaign will last

- What happens to winning ideas or value propositions—how they will move through the system; for example, one innovation architecture might offer a three-step process, where
  - Winning value propositions are funded for proof of concept and developing a business model.
  - Winning business models go on to another round where they get additional funds to build a team, create prototypes, and plan for customer acquisition.
  - Finally, in a third round, one or more projects get funding for pilots and implementation.

Innovators also typically want to know what their own role will be if their idea is selected to move forward. How will they be involved? Can they continue to work on their idea or would they turn it over to others? And of course, they always want to know how they or their team will benefit from their idea's success.

Finally, it is helpful to have an innovation team to help train users, to create in-person or online events and rewards, and to celebrate successes. This team increases user participation by ensuring the process is fun and engaging, that innovators learn something worthwhile from their engagement, that campaign questions are compelling, and that innovators know why their input is vitally important.

SUMMARY

In summary, online innovation management platforms work best when paired with committed innovation leaders who engage with users on the system, an inspiring innovation strategy, a clear and widely understood innovation architecture and practices, and a vibrant innovation culture.

EXAMPLES

Please see examples on pp. 267–273.

SOFTWARE

There are dozens of online innovation management platforms, and the specific features each one offers change frequently as applications are continually upgraded and new features are added. Platforms that offer a comprehensive set of features, along with an intuitive, easy-to-use interface include

- BrightIdea: http://www.brightidea.com
- IdeaScale: http://ideascale.com
- Mindjet/SpigitEngage: http://www.mindjet.com
- Qmarkets: http://www.qmarkets.net

SUGGESTED ADDITIONAL READING

For creating an integrated innovation management system throughout an enterprise or extended innovation ecosystem:

Friedman, L. and Gyr, H. *Creating Your Innovation Blueprint: Assessing Current Capabilities and Building a Roadmap to the Future.* In: P. Gupta and B. Trusko, eds. *The Global Handbook of Innovation Science.* New York: McGraw-Hill Education, 2014.

For creating shared strategy throughout an enterprise:

Friedman, L. and Gyorffy, L. *Leading Innovation: Ten Essential Roles for Harnessing the Creative Talent of Your Enterprise.* In: P. Gupta and B. Trusko, eds. *The Global Handbook of Innovation Science.* New York: McGraw-Hill Education, 2014.

Friedman, L. and Gyr, H. *The Dynamic Enterprise: Tools for Turning Chaos into Strategy and Strategy into Action.* San Francisco: Jossey-Bass Business and Management Series, John Wiley, 1998.

Gliedman, C., Burris, P., and Wang, N. The Forrester Wave™: Innovation management tools, Q3 2013: The 14 providers that matter most and how they stack up. Private report available through Forrester Research Service at www.forrester.com.

Gyorffy, L. and Friedman, L. *Creating Value with CO-STAR: An Innovation Tool for Perfecting and Pitching Your Brilliant Idea.* Palo Alto, CA: Enterprise Development Group Inc. Publishing, 2012.

21

Open Innovation Platforms

Maria B. Thompson

CONTENTS

Coming together is a beginning; keeping together is progress; working together is success.

Henry Ford

DEFINITION

Open Innovation is the use and application of collective intelligence to produce a creative solution to a challenging problem, as well as to organize large amounts of data and information. The term refers to the use of both inflows and outflows of knowledge to improve internal innovation and expand the markets for external exploitation of innovation. The central idea behind Open Innovation is that, in a world of widely-distributed knowledge, companies cannot afford to rely entirely on their own research, but should instead buy or license processes or inventions (i.e., patents) from other companies.

USER

This tool should be used by cross-disciplinary teams of diverse experts to yield the best results.

OFTEN USED IN THE FOLLOWING PHASES OF THE INNOVATIVE PROCESS

The following are the seven phases of the innovation cycle. An X after the phase name indicates that the tool/methodology is used during that specific phase.

- Creation phase X
- Value proposition phase X
- Resourcing phase
- Documentation phase
- Production phase
- Sales/delivery phase
- Performance analysis phase

TOOL ACTIVITY BY PHASE

Open Innovation is a tool designed to increase creativity and to stimulate sharing of ideas and concepts in order to define a better way or a new way of operating, or to define a new and better product or service. It is therefore most effectively used during the creation phase and the value proposition phase.

HOW TO USE THE TOOL

In this chapter, we will cover methods for successfully defining and hosting both an external Open Innovation challenge and an internal Open Innovation challenge.

Open Innovation makes use of external ideas and technologies to enhance the enterprise's internal technology base; reduce the cost of research and development (R&D) and time to market; and achieve superior product, service, or process innovations. At the same time, unused intellectual property and technology—latent internal intellectual capital—is made available for other firms to license and use.

We no longer operate in the Ford era, when a lone innovator can independently solve complex problems. Reliance on closed (or internally-sourced) innovation processes no longer makes sense. In this postmodern era, sustainable and differentiated innovation requires commitment from diverse actors for an organization's survival. Open Innovation is an enabling methodology to capture the collective knowledge of diverse subject matter experts (SMEs) to generate a higher quantity of creative solutions to challenging problems. The diversity of perspectives and the complexity of challenges hosted with the Open Innovation methodology, leveraging social networking and crowdsourcing platforms foster both survivability and sustainability in corporations. Collective effectiveness is increased, reducing overall R&D costs. Structured networking and problem solving with a common focus lead to collective learning. Risk is not only held by a small organization of players, but can be shared across internal and external entities, resulting in an increased likelihood of new product development success. A competitive advantage is gained through the intersection of diverse sets of knowledge.

Success criteria should be shared with participants in advance and referenced in idea submission forms to ensure all submitters address them. Detailed descriptions of submitted ideas should enable a thorough cost–benefit analysis in order to calculate the idea's unique value proposition versus other top submissions.

Twelve Steps to Follow

The following 12 steps will ensure a successful internal or external Open Innovation challenge/competition/campaign.

1. Sponsorship

Especially for internal Open Innovation, it is critical that participants know that there is executive sponsorship for the challenge. In the author's experience with internal Open Innovation challenges, employees have many priorities, and the challenge needs to be seen as a priority for the organization and senior management. The best way to get active participation from most employees is to have an executive sponsor send the invitation to their staff about the nature and importance of the problem to the company's customers and business strategy. In addition, the executive sponsor should be an active participant during the challenge, and be the one who personally rewards or recognizes the winning submissions when the competition ends. It is also likely that funding for the challenge, the Open Innovation platform subscription (see section Software), and rewards or intellectual property filings should largely come from this executive sponsor.

For successful diversity of thought, while the original problem may be selected by a primary executive sponsor, most challenges produce a wider variety of unique solutions when secondary sponsors from other organizations with contributing subject matter expertise are engaged. These sponsors can all invite the employees in their organizations to submit ideas, and regular reports about the engagement and quality of submissions from their respective organizations can be provided by moderators (see step 10) using most software platforms.

For external Open Innovation challenges, the engagement of participants is a marketing issue, and using online social media and crowdsourcing tools to target messaging to the most likely communities (e.g., universities, technical consortiums, or independent inventors) with subject matter expertise about the problem domain and commensurate rewards is a must.

2. Problem identification

There are several characteristics that identify a *good* problem for an Open Innovation challenge. It must be a complex enough problem that a couple of SMEs cannot easily solve it on their own. Even if you are going for a design-around solution to a competitor's patented technology, it is important that the problem itself can benefit from the labor and resources consumed by a typical challenge. There are costs associated with the platform you choose to host the challenge and manage submissions, but these are the least of the costs involved. The problem must be important enough that the employees and submitters are allowed a certain percentage of their time away from their regular duties and can brag about their participation without fear of retribution.

A senior manager in the corporation should identify a technical problem as a potential threat or disruptor to the business. Other sources of problems are long-standing customer issues for which neither your company nor your competitors have elegant solutions. These problem domains will benefit from the efforts to launch and manage the challenge, as well as the efforts by innovators to engage and work on solutions to the problem.

3. Selection of SMEs

Open Innovation, either with internal or external experts, can be used to enhance an enterprise's internal technical base and reduce the cost of R&D, as well as cycle time or time to market. It can result in the creation of superior product and process innovations by bringing together diverse perspectives to build successful solutions for complex problems. It is essential that a core team of SMEs in the specific problem domain is engaged from the start to design the challenge, and to create an engaging tutorial on the importance and impact of solving the high-value problem.

Also, SMEs should engage throughout the duration of the challenge to build on submitted ideas and redirect those that are promising or partial solutions. If the problem is a customer issue, product managers or field service workers who interact the most with the customer on the issue should definitely be part of the SME team to clearly design the content of the challenge, as well as evaluate submissions.

4. Problem reframing and tutorial development

External Open Innovation. Many software platform vendors offer assistance with challenge formulation as part of their subscription fees. Sponsors should define the challenge type, though most Open Innovation challenges are focused on seeking solutions to complex problems. However, in some cases, a corporation may be seeking competencies or talent to partner with, to bring a new product idea to market. In this case, participation by internal SMEs or employees may not be necessary. Also, depending on how constrained the solution space is, external Open Innovation challenges may not need to invest time in problem reframing.

The following outline should be used in the Challenge description, as well as solution requirements, the latter of which should be spelled out in the Idea Submission form:

- The title of the challenge—use action-oriented impactful adjectives and nouns describing the desired end result of the solution.
- History and background of the problem domain
  - Include stories and quotes about the issue from the user or customer, as well as reframe any jargon in generic functional terms a diversity of participants without subject matter expertise can understand (see Domb, 2014).
  - Details of who, what, when, where, why, and how the problem occurs and various contributors (e.g., causes and effects); use cases and scenarios are compelling.
- Solution requirements specifications—specific performance parameters that must be achieved; be sure to provide ranges of performance versus exact values.
- Conceptual directions to steer clear of, due to inherent issues, lack of competency or expertise of implementing workforce, or existing intellectual property.
- Points that should be addressed in the submission; it is critical to ask questions that will draw out the submitter to get them to think through all the possible ways to solve the problem, and all the details that might be relevant. Also consider how you might want to cluster or categorize solutions, so that it is easier to evaluate and organize all submissions. Categories help divide responsibilities among SMEs and moderators, as well as for reporting purposes if one category of solution is not addressed by any submissions—for planning follow-up challenges.
- Duration of the challenge or how long submissions will be taken, so participants can plan their time accordingly. Some of the best results come from localized in-person brainstorming around the problem, reviewing and combining similar ideas, and then asking the participants to submit the top solutions to the online challenge for others to build on (see Seeding in step 8). This is an effective approach when you have geographically-distributed teams of participants.
- Selection and scoring criteria, for example, projected performance parameters that demonstrate the effectiveness of the solution in solving the problem, projected cost of components and

implementation resources, ease of implementation, installation and configuration approach at customer site, etc.

5. Pre-selection of participants

Just because you create a challenge does not mean that people or experts with information and knowledge to share will participate. This is *not* a Field of Dreams situation ("If you build it, they will come"). It will be important to engage the right diversity of expert thought from relevant disciplines. Be sure to clearly define the WIIFT (*what's in it for them*) to participate. See Rewards and Recognition (step 11) and Sponsorship (step 1). Be sure that the organizations from which you want the most participation and thought leadership for internal Open Innovation have strong management sponsorship, and have been invited by their management to participate. We have found that sending personalized e-mails from the sponsor's account to all anticipated participants is the best approach. SMEs and innovation champions can ghostwrite these email invitations and assemble the appropriate distribution list of diverse experts.

6. Question Banking

Once you have an original tutorial available from the SMEs that describes the impact of the problem when it occurs, what causes it to occur, and what the causes and effects are, it is good to chunk the various aspects of the problem down into thought-provoking open-ended questions that participants can engage with. One of the best approaches for identifying questions to ask is by performing a function analysis on the main problem, and then identifying all the components and tasks that contribute to the problem and how they function with action verbs to contribute. Once you have this list, you can ask open-ended questions about how each function or task might be performed differently to avoid the resulting problem. Studies have shown that asking questions stimulates more brain activity (see http://www.bartelart.com/arted/questions.html, Bartel, 2004).

7. Duration

Internal. The duration of most challenges should not exceed two to three weeks. It is best if the participants are not in the middle of major deliverables for internal Open Innovation challenges. Sometimes, it may be necessary to have challenges run longer during a holiday break; however, regular reminders about the challenge completion date should be sent out to all participants. In almost all internal competitions the author has participated in, the most submissions came in right before the deadline. In these cases, you may need to have follow-up meetings with top submitters and SMEs in order to derive complete and robust solutions.

External. Durations are determined by the sponsor. It is assumed that the sponsoring company is seeking external expertise to solve a problem or generate an idea because they do not have internal expertise in the area. For this reason, challenges typically have longer durations of one to three months, and when fully functioning prototypes are sought from submitters, the duration can be from three to four months. Most often the moderators are not internal employees in these types of challenges. More sharing occurs with the Open Innovation platform vendor, who moderates and prescreens the submissions before bringing the top contenders to the attention of the corporate sponsor.

8. Seed ideas

For internal Open Innovation challenges, it is important to seed the competition with at least partial solutions and include some questions or discussions of the value of the respective seed ideas, or how they might better address the problem. Once you get the ball rolling by having SMEs contribute their best ideas, others will be inspired to also submit theirs. In introverted cultures, it may be best to conduct brainstorming sessions where people work individually and together to generate ideas in a face-to-face collaboration, and then preview the best ideas and ask those with the best ideas to submit them online. This way, the more junior or introverted participants will know that a panel of experts feel their idea has value and is worth submitting to the challenge, when otherwise they would have opted out of participating.

Building on others' ideas will still get credit to individual contributors on most platforms, since every participant's employee ID is tracked for each comment or idea. This way, if you decide to patent any final solution that several people contributed to, all contributors can be engaged in the patenting process.

9. Security mechanisms

External. Many sponsors do not wish to reveal their identity to external submitters. This helps them protect their intellectual property and prevents their competitors from understanding their areas of need or difficulty. Confidentiality can be maintained by reframing specific problems generically. Also, by not being specific about the industry of the problem (see Reframing in step 4), a more diverse set of problem solvers can be engaged to submit novel solutions.

Internal. Security mechanisms for internal Open Innovation challenges are primarily with regard to privacy of data stored on an external software platform. Corporations should ensure the software platform is protected from access by other corporations. See the software platform requirements below.

However, in some cases, large corporations may want to sponsor private challenges and only provide access to those sponsors and employees specifically invited to participate in the challenge, due to technical constraints of the solution space and knowledge or expertise required by participants. Also, it may be important to prevent consultants or contract employees from viewing the ideas submitted in order to avoid intellectual property rights contamination.

10. Moderators

See step 3 above, and engage SMEs or innovation champions who work closely with SMEs to moderate and provide timely feedback, build on, and evaluate solutions submitted by participants. You will find that many submissions will consist of only partial solutions, and moderators will need to ask clarifying or building questions in order to help submitters derive complete robust solutions for the problem. Often, the participants with the least experience in the problem domain will contribute partial creative solutions that will spark thoughts among SMEs who might be too close to the problem domain or have been working on the problem for a very long time.

We cannot solve our problems with the same thinking we used when we created them.

Albert Einstein

Furthermore, moderators will be able to readily score or evaluate ideas they have seen or tried before and can aid the submitter in redirecting their original solution approach in a more fruitful direction. If SMEs are moderators, they will also be able to identify solutions that are already patented by the competition and must be eliminated or redirected to a design-around solution.

Moderators should also ensure the advertisement of top participants and top-scored ideas throughout the competition. Most Open Innovation platforms allow participants to score each other's submissions. Another common approach is to let various SMEs score the submissions live, so that other participants see the top ideas bubbling up and can contribute and strengthen these ideas if their solution is similar.

11. Rewards and recognition (including leader board)

External. For external Open Innovation, recognition can take the form of lump-sum awards, intellectual property licensing and commensurate royalty payments, or purchase of patent rights. All submitters typically grant the challenge sponsor a non-exclusive license to use their submissions as part of the challenge registration process. Most corporations would not willingly conduct external Open Innovation challenges unless there is assurance of intellectual property ownership for their own freedom to use the selected/winning solution. Sponsors should also designate whether the winning submitter can work on a solution anywhere in the world or in a specific location. For challenges in which the reward is a cash prize, the sponsor can grant partial awards for solutions that do not address all the requirements. Also, the sponsor can invite selected winners to job interviews or establish a joint venture. The challenge sponsor can specify the type of relationship they are considering with the winning submitter, that is, whether they are interested in employing new talent, a consulting relationship, a potential internship, partnership, joint development, supplier agreement, or other.

External challenges can be written in a way that does not reveal confidential information or the identity of the sponsor. This can be done by describing only the specific problem that needs to be solved without identifying the general industry domain of the problem.

Sometimes, a challenge can serve the purpose of tech scouting when the experts that engage derive creative and differentiated solutions for which the sponsoring corporation does not possess knowledge or competency. In this scenario, the challenge can also result in a joint venture, acquisition, or hiring agreement. These options can result in reduced cycle time to market, and legal advice should be sought early in the challenge cycle to ensure intellectual property rights are clearly understood by the sponsor and the participants. It also helps for the sponsoring organization to perform a patent landscape analysis before advertising the challenge, in order to steer participants away from patented solutions.

Higher awards will result in greater interest from submitters. It is possible that submitters may only partially fulfill the challenge solution requirements but still provide valuable ideas. To reward these submitters, other recognition awards may be given.

Internal. Often, submitters to internal Open Innovation challenges seek rewards that are nonmonetary in nature. Recognition from senior management, extra days off work, designated parking places, special trips, or gift certificates are all welcomed by employees, especially public recognition to all participants, including the reason the selection was the best of the competition.

Publicly acknowledging the individual submission and all the people who contributed to the idea and participated in the challenge, as well as those who commented and built on others' ideas, goes a long way to creating a corporate culture of crowdsourcing. Friendly internal competition for top spots on the Open Innovation leader board that senior management regularly sees, discusses, and recognizes with personalized communications motivates engagement in future challenges.

12. Aggregation, combination, evaluation

Success criteria to evaluate proposed solutions should be clearly defined, advertised as part of the invitation to the challenge, and included in the online tutorial for the problem domain. Then, it can be used to evaluate whether a proposed solution contributes to the sustainability and differentiation of a company's product portfolio.

During the competition, moderators and SMEs can perform ongoing evaluation and scoring of submissions in order to get timely feedback to submitters. Sometimes, combining two submissions will result in inspiring additional ways to build on the original ideas to create a more robust solution. It is important to quickly combine or retire solutions that do not meet criteria, so submitters get timely feedback and expend efforts on the most useful solutions and concepts. Sometimes, the best ideas come from multiple submitters building on another submitter's original idea.

EXAMPLES

Internal

Illinois Tool Works (http://itw.com) investigated Open Innovation to

- Create connection and collaboration throughout ITW's seven diverse industrial segments to leverage diversity of thought

- Enable generation of innovative solutions addressing important business needs

At ITW, internal Open Innovation challenges can enable rapid development of actionable solutions to difficult problems by engaging multiple diverse solvers.

Defined success criteria for challenges include

- Engage participation from multiple ITW segments/divisions
- Generate viable solutions
- Frame the right problem(s)
- Clearly define problem in generic terms (stripped of domain-specific jargon)
- Solution success criteria for addressing the business need
- Executive sponsorship
- Collaborative portal to create a conversation with innovators across the company
- Clear direction on how submitted solutions will be funded or implemented
- Recognition for innovators by senior leadership

Motorola Solutions Inc. hosted several internal global Open Innovation challenges on the Bright Idea Open Innovation platform to successfully engage diverse populations from different cultures and countries around the world.

External

General Electric

General Electric is a role model for launching several successful external Open Innovation challenges on a variety of platforms. See their Open Innovation Manifesto at http://www.ge.com/about-us/openinnovation. An overview of why GE felt they could use Open Innovation as a strategic advantage is described here at http://www.wired.com/2014/04/how-ge-plans-to-act-like-a-startup-and-crowdsource-great-ideas/.

Several challenges that GE has successfully launched are described at http://www.geglobalresearch.com/impact/open-collaboration-companies-big-small.

IBM

IBM has been successfully using both internal and external Open Innovation since 2001. They have an internal platform that all employees contribute to, and can see and build on each other's ideas, called Thinkplace (http://www.slideshare.net/kapilgupta/think-placeoverview-external20084q).

Idea connection has also hosted successful external Open Innovation challenges for IBM, in addition to other software vendors listed above. See one example at http://www.ideaconnection.com/open-innovation-success/Big-Green-Innovations-00147.html.

Procter & Gamble

Many successful stories of how P&G has hosted external Open Innovation challenges are available at http://www.pgconnectdevelop.com/home/stories.html. Important to note about the history P&G has with Open Innovation is that, while they were one of the first corporations to embrace using external experts for problem solving and new product development, they also have found that their internal competency in innovation and creative problem solving has waned. This should be a wake-up call to any corporation considering Open Innovation as the *only* way to consistently generate new products and solutions, and some of the prior references about using challenges can help corporations determine potential new hire candidates, joint venture candidates, as well as acquisition candidates to bolster their internal knowledge-worker population as a competitive strategy.

SOFTWARE

Some commercial software available includes, but is not limited to

- Idea Connection: http://www.ideaconnection.com/software/
- Bright Idea: http://www.brightidea.com/
- Innocentive: http://www.innocentive.com/
- Nine Sigma: http://www.ninesigma.com/
- Cognistreamer: http://www.cognistreamer.com/
- Mindjet/Spigit: http://www.spigit.com/
- We Bridge Innovation: http://webridgeinnovation.com/

See Table 21.1.

TABLE 21.1

Considerations for Selection of an Open Innovation Software Platform

Attribute	Description	Comments
Access control mechanism	Administrative capabilities to set access control privileges by user, group, and roles for challenges/campaigns, postings, reports, and printouts (Read, Write, Modify, Full Control)	Security: Need to protect intellectual property from external clients with firewalls, as well as between organizations and roles within the system
Reporting capabilities	Ability to construct and build reports on any data/fields stored in the system; all reports and fields must easily export to Excel format	Important to easily create dashboards and reports for management and different groups
Tagging or meta data capabilities (default and user defined/challenge sensitive)	Allow for users or administrators to assign keywords to each posting for easy indexing, searching, retrieval, and reporting	Suggested meta tags: technology area, business/division, role, ownership, product, date
Role definition, assignment	Roles are associated with states in the workflow in the system, and have specialized access privileges	Workflow and access privileges are role based and roles can be defined, associated with fields, and actions on those fields by the ITW administrators and moderators
Automated workflow	System should support idea management such that as meta-data of the original posting are updated, the idea is automatically made visible and assigned to designated roles for specific action	This automated workflow capability must be easily tailored to align with the size and structure of the using organization
WYSIWYG (*what you see is what you get*) editing	Ability to make the site, challenges, postings, images, fields, categories, etc., look and feel like you want	No need to involve IT or pay for additional consulting/programming by the vendor to change the look and feel of the user interface or workflow or access privileges or reports

(Continued)

TABLE 21.1 (CONTINUED)

Considerations for Selection of an Open Innovation Software Platform

Attribute	Description	Comments
Integration with document management system	Should play nicely with SharePoint, Documentum, Autonomy, or other documentation management systems	Rather than having submitters upload large documents into their idea forms, they should be able to link to a secure folder on a document management site where they can upload all related documents and restrict access privileges. This document management system should index all documents from submitters for full-text searching by moderators, who can track whether similar ideas have already been submitted
Full-text searching capabilities	All contents are indexed, with a search engine that allows for Google-type or Boolean searching across all fields in the system, depending on access privileges (administrators have all access)	Sorting and searching is default mechanism for tracking and finding information in the system. Ability to find *more like this* once have a result; list of all text of all postings
Non-proprietary, noncustom data based on back-end	SQL server or whatever IT or vendor has experience with and can perform database administration on—DBA should be part of service contract	Allow for views and integration with other systems through SQL connectors or similar technology
Forms creation, management	Ability to design tailored forms for idea submission, automated e-mail alerts, notifications, etc.	
Leadership boards	Automated tracking, aggregation, and display of counts of most submissions, most comments or builds, most likes, etc.	Allow for leadership boards, displays, reporting, recognition
Easy login, floating concurrent usage licensing	No restricted, single-named user licensing—not a good model for ITW's decentralized culture	Good for one click registration with initial password e-mailed for easy log-in… administrative override privileges for ITW admins to clear passwords for user reset
Service Level Agreement	Contract should stipulate response times, geographic coverage, outage support/hotline, and reduction in costs if system is down or has issues	Also, regular maintenance in compliance with our records retention policy should be performed (by tagging/data types and purge schedule we stipulate)

REFERENCES

Bartel, M. Teaching creative thinking with awareness and discovery questions (inquiry learning). Available at http://www.bartelart.com/arted/questions.html, 2004.

Domb, E. *Global Innovation Science Handbook.* Function Analysis as discussed in Chapter 24—TRIZ: Theory of Solving Inventive Problems. New York: McGraw Hill Education, 2014.

SUGGESTED ADDITIONAL READING

Benner, M.J. and Tushman, M. *Process Management and Technological Innovation: A Longitudinal Study of the Photography and Paint Industries.* Open Innovation. Cambridge, MA: Harvard Business School Press, 2002. Cited by 326 related articles.

Caporaso, J.G., Kuczynski, J., Stombaugh, J., and Bittinger, K. QIIME allows analysis of high-throughput community sequencing data. *Nature,* 2010. nature.com.

Chesbrough, H.W. and Garman, A.R. How open innovation can help you cope in lean times. *Harvard Business Review,* 2009. europepmc.org. Cited by 142 related articles.

Dittrich, K. and Duysters, G. Networking as a means to strategy change: The case of open innovation in mobile telephony. *Journal of Product Innovation Management,* 2007. Wiley Online Library. *Strategic Management Journal,* vol. 22, no. 6–7, pp. 521–43.

Elmquist, M., Fredberg, T., and Ollila, S. Exploring the field of open innovation. *Journal of Innovation,* 2009. emeraldinsight.com. The study resulted in 49 publications, of which 35 are journal articles, 10 are book reviews or columns, and 4 are books. The book reviews and columns discussing or mentioning open innovation in general terms were excluded from further analysis of the publications. Cited by 162 related articles.

Enkel, E., Gassmann, O., and Chesbrough, H. Open R&D and open innovation: Exploring the phenomenon. *R&D Management,* 2009. Wiley Online Library. Cited by 653 related articles.

Gassmann, O., Enkel, E., and Chesbrough, H. The future of open innovation. *R&D Management,* 2010. Wiley Online Library. Cited by 550 related articles.

ISME Journal original article: Open Innovation Challenges. Boehringer Ingelheim challenge: Understanding the antidepressant effect of ketamine. Cited by 2623 related articles.

Keupp, M.M. and Gassmann, O. Determinants and archetype users of open innovation. *R&D Management,* 2009. Wiley Online Library. Article first published online: August 4, 2009. doi: 10.1111/j.1467-9310.2009.00563.x. Journal compilation. © 2009 Blackwell Publishing Ltd. Issue. R&D Management. Special Issue: Open R&D and Open Innovation Edited by Ellen Enkel and Oliver Gassmann. Cited by 185 related articles.

Lichtenthaler, U. and Lichtenthaler, E. A capability-based framework for open innovation: Complementing absorptive capacity. *Journal of Management Studies,* vol. 46, no. 8, pp. 1315–1338, December 2009. Article first published online: July 6, 2009. doi: 10.1111/j.1467-6486.2009.00854.x. © Blackwell Publishing Ltd, 2009. Available at http://onlinelibrary.wiley.com/doi/10.1111/j.1467-6486.2009.00854.x/abstract ;jsessionid=7465CC0287866F55259C44A9013CCBEC.f04t04?userIsAuthenticated =false&deniedAccessCustomisedMessage=.

OIC Editor. Feds developing plan for open innovation toolkit. Available at http://www.open innovation.net/open-innovation/feds-developing-plan-for-open-innovation-toolkit/.

OIC Editor. The value of open innovation for B2B companies. Available at http://www.openinno vation.net/open-innovation/the-value-of-open-innovation-for-b2b-companies-2/.

Spithoven, A., Clarysse, B., and Knockaert, M. Building absorptive capacity to organize inbound open innovation in traditional industries. *Technovation*, 2011. Elsevier. In line with Cohen and Levinthal's seminal article, absorptive capacity is usually opera-tionalized as the existence and/or intensity of a company's R&D activities (Veugelers, 1997; Lane and Lubatkin, 1998). Inbound open innovation and absorptive capacity in traditional industries. Cited by 310 related articles.

22

Outcome-Driven Innovation

Dana J. Landry

CONTENTS

DEFINITION

Outcome-driven innovation is built around the theory that people buy products and services to complete tasks or jobs they value. As people complete these jobs, they have certain measurable outcomes that they are attempting to achieve. It links a company's value creation activities to customer-defined metrics. Included in this method is the opportunity algorithm, which helps designers determine the needs that satisfied customers have. This helps determine which features are most important to work on. Most important is this tool's intention of trying to find unmet needs that may lead to new and innovative products/services.

USERS

This tool is best used by small teams focused on creating new or significantly improved products or services. This tool is best used by cross-functional teams.

OFTEN USED IN THE FOLLOWING PHASES OF THE INNOVATIVE PROCESS

The following are the seven phases of the innovative cycle. An X after the phase name indicates that the tool/methodology is used during that specific phase.

- Creation phase X
- Value proposition phase
- Resourcing phase
- Documentation phase
- Production phase
- Sales/delivery phase
- Performance analysis phase

TOOL ACTIVITY BY PHASE

- Creation phase—This process is directed at collecting information related to the customers' unfulfilled needs and helping the designer choose the ideas that best address the customer's unmet needs.

HOW TO USE THE TOOL

- This tool requires significant customer interaction through surveys that themselves require expert question development.
- This tool requires training from competent sources to be best applied.
- This technique is most successfully applied through a carefully considered approach with a clear goal for why the tool is being used and what it is trying to accomplish.

EXAMPLES

The output of this tool is generally a graphic that shows a continuum from unmet to well met need against importance. The intent of the graphic is to show the most important unmet needs. These are then considered to be the best source for new opportunity.

REFERENCE

Ulwick, A. *What Customers Want: Using Outcome-Driven Innovation to Create Breakthrough Products and Services.* McGraw-Hill, New York, 2005.

SUGGESTED ADDITIONAL READING

Ulwick, T. PowerPoint presentation downloaded from seanmiller.blogs.com/whizdumb /files/TonyUlwick10182007.ppt.

23

Proactive Creativity

H. James Harrington

CONTENTS

Everyone is creative but the world is trying to break us of that habit.

H. James Harrington

DEFINITION

Proactive creativity is a process that places the individual into an environment or a mental state that promotes right-brain thinking.

USER

This tool is best used by an individual in a private environment, although it can be used when exercising with a large group of people performing the same exercise.

OFTEN USED IN THE FOLLOWING PHASES OF THE INNOVATIVE PROCESS

The following are the seven phases of the innovative cycle. An X after the phase name indicates that the tool/methodology is used during that specific phase.

- Creation phase X
- Value proposition phase X
- Resourcing phase X
- Documentation phase X
- Production phase X
- Sales/delivery phase X
- Performance analysis phase

TOOL ACTIVITY BY PHASE

This is a tool that is used to remove an individual from the day-to-day stress, allowing them to relax and let their mind to wander into new thought patterns. Although it can be used in most of the phases to help solve difficult problems, it is best used in phases that are not strictly controlled or when the results of the activity are not predefined. Examples include research and engineering, sales and marketing, and building design.

HOW TO USE THE TOOL

Far too often, people only call on their creative powers when they are faced with a problem. This is truly unfortunate because they underutilize this gift by not applying it fully. As a result, they develop a reactive rather than a proactive approach to creation. In reality, we believe that individuals need to develop and use both their proactive and reactive creative powers in order to make maximum use of their creative potential.

There are different reasons that drive individuals or groups to become creative. The most common are

- A significant emotional event or a traumatic emotional event—For example, solving a problem: a car that will not start in the morning requires us to create a new way to get to work.
- Playful brainstorming—Listing of new ways to come up with something; for example, a new way to serve hot dog, like serving hot dog on a stick.
- Systematic creativity that is purposeful—The goal is to fill a void or come up with a better way to do something. It need not be playful or problem solving in nature.
- To satisfy a personal desire—Some individuals are driven to look at things in a different way, or they feel the personal need to be creative.

To sharpen your creative powers, you need to establish a workout program, and exercise your creative self at least three times a week. Individual workout sessions can vary from 5 to 60 minutes, depending on the exercises you select and your personal interest in the outcome.

SETTING THE STAGE FOR CREATIVITY

Creativity can occur at any time and in any place. Sometimes, we are very creative. Other times, it is just impossible to pluck out an original thought. I have sat in front of my computer for hours pecking at one key at a time to capture a few unimportant thoughts that I erased at the end of the session. At other times, the ideas flow out of my mind so rapidly that I lose them because I cannot record them fast enough. There is a lot that we can do to prepare ourselves to become more creative. We can position and train the right (Oscar) side of our brain to speak out more often, and the left (Felix) side of our brain to listen more intently and openly to Oscar's ideas. For this to occur, three conditions need to be present (Harrington et al., 1998):

- **Time.** Extra time is often required to develop and sell a creative solution that is not in line with the individual's or organization's culture.
- **Environment.** It is difficult to be truly creative when you are continuously being interrupted with phone calls, questions, or have children climbing on your lap.
- **Success.** Nothing gets Felix's attention better than when we are recognized because we come up with creative new solutions.

Creativity Workspace

Our emotions and actions are directed by our preconceived notions about the environment we find ourselves in. We enter the library, and we talk softly and move carefully. We go to a party, and we laugh and smile more. We pick up a baby, and we coo and gurgle like we have no mind at all. We go to work, and we become more conservative, reserved, and formal. We go to a movie, and we sit beside a friend for hours without talking. This behavior is not only acceptable but expected. We have been trained from birth to conform to the expectations related to the environment in which we find ourselves (Harrington and Lomax, 2000).

We like to set aside a specific location where we exercise our creativity. In Jim's case, it is a desk in a small back bedroom. Before that, it was a credenza behind his desk at work. One of Bob and Glen's favorite places is the beach. They discovered this when they developed an earlier book while they were conducting a class in Curacao. They spent each evening on the beach working on their new book. It does not have to be a grand place. It could be a workbench in the garage or an old desk in the cellar behind the furnace. The important thing is that in your mind and in your family's or business associate's mind, it is your space and there are specific rules associated with it.

Rule 1. No interruptions are tolerated unless it is an emergency.

Rule 2. The clean-desk policy does not apply. Do not take time to organize the work area, and it is out of bounds to your spouse or your coworkers. Remember, the world of Oscar is one of clutter and disorganization. Just think of the time you will save by not having to pick up, put away, and get out the same materials later.

Rule 3. Make your creativity place very visual. Use lots of Post-its. Write down good ideas on them and stick them up around your creativity area. Make sketches and flow diagrams, and put them on the walls.

Put up very different, interesting pictures and change them often. Your creativity place is to stimulate ideas, not to impress others.

Rule 4. Have a relaxed atmosphere. Have a comfortable chair—one that you can lean back in and rest your head while your mind goes blank and opens to creative thoughts. Have furniture that you can put your feet on. Choose a spot that is not too hot or cold.

Rule 5. Have the right equipment. Be prepared to be flooded with new ideas, and when they come, you need to be able to capture them rapidly. Things that can be useful are

- Good lighting
- A computer
- Lots of paper
- Colored markers
- A tape recorder
- A CD or tape player (for restful music)
- A filing system
- A corkboard
- A bookcase

Rule 6. Have a focal point. This is something that relaxes you when you look at it. It could be a window that you look out of, or a small aquarium. Other people find an ocean scene or an abstract painting does the job for them. Use whatever relaxes you.

Each person's creativity place is very unique to them since it must fit into their individual personality and output expectations. We are not telling you that you must have a creativity place, but we strongly suggest that you should consider using a creativity place as a tool to help turn on your creative juices. Does this mean that this is the only place where you will be creative? The answer is a resounding no. It is a lot like the treadmill that you buy and put in your house to jog on. When you get on the treadmill, you do not start eating a sandwich; you start to jog, and because you have a treadmill, it does not mean that you cannot run around the block in front of your house.

Relax to Create

Often we find ourselves caught up in the stress and strain of our everyday activities. Johnny gets hit in the head with a rock on the way to school. Johnson cancels the order. You forgot to send out the new schedule. Alice is sick, and no one knows what she did with the drawings. Your flight to Kent, Nebraska, was canceled. How can you turn off all of this and focus on being more creative when it is all you can do just to tread water and keep things going?

If this describes your situation, it is best to start your creativity session with what a jogger would call a stretching exercise. We call them calming or meltdown exercises. These are exercises designed to prepare the muscles that you are about to use for the workout they are about to undergo. There are a number of these meltdown exercises that are used. We particularly like meditation.

Meditate to Create

Ever since meditation exploded in popularity during the 1970s, devoted followers of all ages have endlessly praised its benefits and how it revolutionized their lives. Many testimonials describe decreased stress levels, increased energy levels, increased productivity, and increased creativity. For example, a study showed that by meditating 20 minutes twice a day, people aged 40 and older have 74% fewer doctor visits and 69% fewer hospital admissions than the average population. Another study at a New England bank and northeastern industrial firms revealed that meditators experienced fewer errors, less absenteeism, more focus on work, and more creative ideas. Finally, researchers discovered that meditation helped business school students solve problems faster and develop more effective teamwork than the other students.

These benefits are so compelling that many businesses, including the Adolph Coors Company, AT&T, New York Telephone Company, and Hoffman-La Roche, have adopted meditation as part of their stress management programs. Not only do they save money, but they also provide the company with a competitive advantage by increasing productivity and creativity in their employees.

With all the hype, many began to wonder what exactly meditation is. Skeptics have the preconceived notion that it is a bizarre New Age religion. It can be confusing and easy to dismiss because there are so many meditation schools, approaches, and techniques. These aspects of meditation become more elusive with the technical jargon used or all the different levels of meditative states.

A universal characteristic of meditation is the elevation to an altered state of consciousness by clearing the mind. Transcendental meditation (TM) rose as the prevalent meditation program after the Beatles publicly endorsed it. Founded by Maharishi Mahesh Yogi, the TM method can only be learned from teachers taught by the Maharishi himself. There are Maharishi Vedic Universities throughout all 50 states where one can become initiated into the TM program. Other meditation approaches include

- Clinically Standardized Meditation (CSM) developed by Dr. Patricia Carrington, author of the well-received *Freedom in Meditation*.
- Respiratory One Method (ROM) developed by Dr. Herbert Benson, author of *The Relaxation Response*.
- Self-hypnosis.

So, why should we meditate? Our lives are so busy and hectic that most of us walk around with many thoughts racing through our minds at the same time. Meditating helps us clear our minds of all those thoughts, and relax so that we can focus on one thing at a time. Imagine piles of paper strewn all over a desktop. If you add another report to the piles, it would be difficult to notice that report. Furthermore, you would not be able to concentrate on anything because your thoughts easily wander to all the other work on your desk. By clearing the desk of all the clutter, it is much easier to focus on a single item.

Concentration on one thing at a time is more effective in developing creative ideas. The most creative geniuses, such as Einstein and Mozart, sustained intense periods of absorption that are similar to states achieved through meditation. An eight-year-old violinist at the Julliard School of Music meditates to relax and be more expressive. We have all experienced such states of absorption while watching a movie or listening to music that we lose sight of our surroundings and the time. That is when our creative potential is the greatest.

Another powerful reason for meditating is that meditation alleviates stress. Not long ago, debates over whether stress induced creativity remained unresolved. However, it is now widely accepted that the opposite is true. You will not be less creative by eliminating stress. In fact, stress can sometimes inhibit creativity.

Try the following CSM exercise. Read all six steps then go back and start to use them in numerical order

Step 1. Go to a quiet place in your house or close the door to your office.

Step 2. Sit upright in a chair with your hands in your lap and your feet placed flat on the floor. It is important to sit with your back perfectly straight to continue the exercise without getting tired.

Step 3. Close your eyes and relax all your muscles.

Step 4. Focus on a calming word (a mantra), phrase, or picture. For example, use the word *relax* or think of the ocean.

Step 5. Repeat the mantra and let it come and go through your mind. Do not try to force it.

Step 6. Continue meditating for 10–20 minutes.

If you have not meditated before, it will be difficult to stay still for so long. You may even experience an uncontrollable itch or restlessness all over your body. But keep trying. It will take a while to build your endurance for meditating.

There are many excellent cassette tapes and disk that can be purchased from any bookstore or music store that will help you relax and increase your mind's susceptibility to new thought patterns. These tapes rely on subliminal learning or self-hypnosis. In some cases, the moderator whose voice is soft and melodic (reminding you of the comfortable feeling you get when you rub velvet against your face) takes the listener on a vacation to a relaxing and quiet spot as he or she describes the place with many picturesque words. Typically, these quiet spots are under a big tree beside a rippling stream with the warm autumn sun shining on your face while billowy white clouds float lazily through the dark blue sky. The moderator encourages the listener to sit back and enjoy the environment and feel the rush of air in and out of his or her lungs. With each breath, the moderator encourages the listener to exhale his or her problems and pressures, and breathe in clean, uncontaminated fresh air. The object is to have the listener acquire a relaxed state of mind by listening intently to a tape for about 10 minutes.

Another type of tape focuses on relaxing individual parts of the body until the body is totally relaxed. The moderator may start by suggesting to you that you relax your jaw, then your eyes, shoulders, fingers, toes, and so on. The moderator will then typically have you focus on the peacefulness of your breathing, your body warmth, and the warm feeling you get when thinking about someone you really care about. Often, these tapes are complemented with soothing music or environmental sounds like the ocean or raindrops. Other tapes use music and environmental sounds as the primary cassette message that contains subliminal messages designed to reduce guilt and anger, and reinforce your confidence and feeling of self-fulfillment.

Two tapes that we particularly like are *Learn to Relax* by Dr. David Illig, published by Metacon Inc., and *Ten Minutes to Relax* based on the book *Relaxation and Stress Reduction Workbook* from New Harbinger Publishing.

Typically, 10–15 sessions, three times a week, are adequate for most people, with additional sessions when the individual has had a particularly bad day. The session will start by the listener finding a quiet spot where he or she will not be interrupted and can close his or her eyes for 10–30 minutes. Then, the listener should loosen his or her clothes before the tape player is turned on.

Of course, meditation is not a panacea. Many have tried meditation and found unsatisfying results. However, you will not know that it is not for you unless you try it first. There are many meditation books available at the bookstore or library. For the busy novice meditator, Salle Merrill Redfield's *The Joy of Meditating: A Beginner's Guide to the Art of Meditation* is a short book that guides the reader through four brief meditation exercises. *Complete Meditation* by Steve Kravette is a thorough collection of exercises with a specific exercise for creative inspiration and invention. *The New Three Minute Meditator* by David Harp provides a humorous approach by demystifying meditation and is excellent for all levels of meditators. Although there are numerous meditation books available, it is best to first seek out a meditation class to attend.

The creative response usually happens in a state of relaxed attention.

Adelaide Bry

Releasing Tension

As we become more and more tense, we rely more and more on the Felix side of our brain to respond to external and internal stimulations. Pressure, tension, worry, and anxiety build a brick wall between Felix and Oscar that is often difficult to break down. We have found that one of the very best ways to disintegrate this obstacle is by covering it with sweat. Physical exercise makes our body healthy, which includes our mind, thereby helping us to be more creative. You do not have to lift weights or do push-ups to increase your creative abilities. There are a lot of less strenuous physical exercises that will increase your mental capabilities. We like the following approaches:

- **Stretching.** Select exercises that stretch your cerebellum, motor cortex, and limbic systems. There are a number of stretching positions that stimulate various portions of the body and brain. Selecting different sections of the brain helps relieve pent-up emotions.
- **T'ai chi.** This is a group of ancient Chinese physical exercises that can be performed by everyone from age 5 to 100 or older. Its slow, rhythmic motions bring the body and mind into complete harmony, reducing and often eliminating stress.
- **Aerobics.** Medical research has proven that there is a direct relationship between cardiovascular health and brain function. Physical exercises like swimming, hiking, walking, dancing, and golf are excellent ways to develop a healthy mind and a creative spirit.
- **Yoga.** Yoga is a longevity mind expander that uses body positions to induce wave patterns in the brain that increase blood flow to the brain, thereby improving its performance.

We recommend that the reader become familiar with each of these four types of brain expanders, and experiment with them to determine which approach or combination of approaches produces the best results for him or her. Once this is accomplished, the reader should design a physical mind expander program so that it concentrates on the optimum combination for the individual while periodically rotating into each of the four approaches. Individuals often find that one combination becomes very effective for a period of time and then another combination takes over. The continuous variation in different physical mind expanders is an important part of a well-rounded program.

> If you choke off the silly ideas and only receive the good ones, the safe, can't miss ones, you might not get any ideas at all.
>
> **Marshall J. Cook**

We have introduced you to mind expanders that are designed to change the way you interface between the Felix and Oscar sides of your brain.

We have shown you ways to relax, thereby helping you prepare to accommodate the creative powers that you already have, and have provided you with mind expanders that will help you think differently. We strongly believe that

- When you enjoy being creative, you will seek out opportunities to be creative.
- When you enjoy what you are doing, you will be more creative.

The only person who can make you more creative is you!

EXAMPLES

Examples are included in the section of this chapter entitled "How to Use the Tool."

SOFTWARE

No software is required for this tool/methodology.

REFERENCES

Harrington, H.J. and Lomax, K. *Performance Improvement Methods*. New York: McGraw-Hill, 2000.

Harrington, H.J., Hoffherr, G., and Reid Jr., R. *The Creative Toolkit*. New York: McGraw-Hill, 1998.

SUGGESTED ADDITIONAL READING

Asaka, T. and Ozeki, K., eds. *Handbook of Quality Tools: The Japanese Approach*. Portland, OR: Productivity Press, 1998.

Brassard, M. *The Memory Jogger Plus*. Milwaukee, WI: ASQ Quality Press, 1989.

24

Proof of Concept

Douglas Nelson

CONTENTS

A picture says more than 1000 words—a model tells the whole story.

DEFINITION

A proof of concept (POC) is a demonstration, the purpose of which is to verify that certain concepts or theories have the potential for real-world application. A POC is a test of an idea made by building a prototype of the application. It is an innovative, scaled-down version of the system you intend to develop. The POC provides evidence that demonstrates that a business model, product, service, system, or idea is feasible and will function as intended.

USER

This tool can be used by individuals, but its best use is with a group of four to eight people. Cross-functional teams often yield the best results from this activity.

OFTEN USED IN THE FOLLOWING PHASES OF THE INNOVATIVE PROCESS

The following are the seven phases of the innovative cycle. An X after the phase name indicates that the tool/methodology is used during that specific phase.

- Creation phase X
- Value proposition phase X
- Resourcing phase X
- Documentation phase X
- Production phase X
- Sales/delivery phase X
- Performance analysis phase

TOOL ACTIVITY BY PHASE

- Creation phase—During the creation phase, the innovator or the creative team will develop value statements that define how the innovative product, service, or situation will affect the customer/receiver of the innovative activity. The POC may be used in refining and validating the value statements.
- Value proposition phase—During the analysis and preparation activities related to preparing a value proposition, the vision of who will be affected and how they will be affected are refined and updated. The goals are also evaluated and updated to reflect the information that has been collected. The POC model is useful in working with potential customers and users to evaluate these value propositions.
- Resourcing phase—The comparison of the resourcing plan and the vision and goals for the innovation concept must be supportive of

each other. The POC is an important requirement in seeking investment from outside parties.

- Documentation phase—As the organization documents the innovation implementation plan and requirements, it must consider the vision and the goals established for the innovation concept. The formal documentation and the vision and goals may be initially validated through the use of POC methodologies.
- Production phase—The goals set related to output performance, costs, and delivery schedules are a primary input in defining how the production activities are organized, and the quality of the production process, and the timing of delivery to external customers. These goals may be further refined and validated through use of POC methodologies such as pilot production.
- Sales/delivery phase—Sales/delivery systems may be refined and validated through POC tools such as implementation of a test marketing program.

HOW TO USE THE TOOL

POC is used within product development.

- Product development:
  The typical stages of new product development are (a) concept development, (b) engineering prototyping, (c) production prototyping, (d) market acceptance testing, and (e) market introduction.
- Proof of concept:
  The steps in POC are (a) definition of success criteria, (b) engineering of a proposed solution, (c) evaluation of proposed solution against success criteria, and (d) decision to proceed.

Stages in Product Development

The product delivery process, as developed by Xerox, identified eight stages in the life of a developing product:

1. Preconcert: An idea for a new product is turned into an outline specification.

2. Concept: The specification is matched against requirements and is defined.
3. Design: The engineers design the product to match the specification provided.
4. Demonstration: The building of prototypes to ascertain correct functioning, followed by pilot production to remove any pilot production problems.
5. Production: Manufacture starts.
6. Launch: Marketing and sales start.
7. Maintenance: Service requirements throughout product life.
8. Retirement: Replacement by a new model.

Other stage gated approaches may divide the product development program more simply as

- Concept development
- Design and development
- Design validation
- Production process development

Prototype

A prototype is an early sample, model, or release of a product built to test a concept or process, or to act as a thing to be replicated or learned from. Prototypes are useful in answering questions about customer reactions, industrial design, fit and finish, and cost of manufacture.

Rapid Prototyping

Rapid prototyping is a group of techniques used to quickly fabricate a scale model of a physical part or assembly using three-dimensional (3D) computer-aided design (CAD) data. Construction of the part or assembly is usually done using 3D printing or *additive layer manufacturing* technology.

Rapid prototyping decreases development time by allowing corrections to a product to be made early in the process. By giving engineering, manufacturing, marketing, and purchasing a look at the product early in the design process, mistakes can be corrected and changes can be made while they are still inexpensive. Rapid prototyping facilitates

- Increasing number of variants of products
- Increasing product complexity
- Decreasing product lifetime before obsolescence
- Decreasing delivery time

Pilot

A pilot (or trial) uses the full production system and tests it against a subset of the general intended market base. The reason for doing a pilot is to get a better understanding of how the product will be used in the field, and to refine the product or service. A pilot is intended to be production quality, although on a smaller scale. Previously completed POCs and prototypes are helpful in validating our approach, minimizing risk, and determining appropriate scope and effort. The pilot is ultimately useful in getting real user feedback.

Test Market

A test market is a geographic region or demographic group used to gauge the viability of a product or service in the mass market before a wide scale roll-out. Criteria used to judge the acceptability of a test market region or group include (a) a population that is demographically similar to the proposed target market, and (b) relative isolation from densely populated media markets so that advertising and promotion within the test audience can be efficient, economical, and measurable.

EXAMPLES

A computer was designed to load a specific set of programs within 30 seconds plus or minus 5 seconds. A test of 50 units indicated that the minimum time was 21 seconds and the maximum time was 27 seconds. Analysis of the data indicated that the maximum loading time met design specifications, but the minimal loading time did not. Analysis of the data indicated that the computer design failed to meet specifications, but the design was accepted because the failure occurred in a positive direction that customers would be more pleased with. The result of this analysis was a change in the engineering specifications to read, "the maximum loading time for the specified programs will be 30 seconds." The reason the specification was changed was threefold:

1. The other functions of the computer would not be affected if the programs were loaded faster.
2. The end-user would be more satisfied with quicker loading times.
3. It costs less to change this back than to redesign the computer.

SOFTWARE

Some commercial software available includes but is not limited to

CAD software tools:
- AutoCAD, Autodesk Inc.: http://www.autodesk.com
- SolidWorks, Dassault Systemes: http://www.solidworks.com

SUGGESTED ADDITIONAL READING

Gebhardt, A. *Rapid Prototyping.* Cincinnati, OH: Hanser Garner Publications, 2003.
Thompson, R. *Manufacturing Processes for Design Professionals.* New York: Thames & Hudson, 2007.
Thompson, R. *Prototyping and Low-Volume Production (The Manufacturing Guides).* New York: Thames & Hudson, 2011.

25

Quickscore Creativity Test

Frank Voehl

CONTENTS

DEFINITION

The Quickscore creativity test is a 3- to 10-minute test that helps assess and develop business creativity and innovation skills.

USER

This tool can be used by individuals or groups, but its best use is with a group of three to six people. Cross-functional teams often yield the best results from this activity, especially when used to gauge organizational innovation capacity.

OFTEN USED IN THE FOLLOWING PHASES OF THE INNOVATIVE PROCESS

The following are the seven phases of the innovative cycle. An X after the phase name indicates that the tool/methodology is used during that specific phase.

- Creation phase X
- Value proposition phase X
- Resourcing phase
- Documentation phase X
- Production phase X
- Sales/delivery phase X
- Performance analysis phase X

TOOL ACTIVITY BY PHASE

- Creation phase—During this phase, the tool is used to determine the current status of creativity and innovation.
- Value proposition, documentation, production, sales/delivery, and performance analysis phases—During these phases, the tool can be used to establish initial creativity and innovation level. It also can be used to measure improvements in creativity and innovation after the initial measurement has been recorded.

HOW TO USE THE TOOL

The Quickscore creativity surveys are a combination of Innovation Coach's two Innovation Surveys and the Next Generation Creativity Survey developed in 2012 by the Centers for Research on Creativity (CRoC). The best way to assess how far you are in creating and sustaining innovation is to first do an innovation evaluation or audit. No matter if you are an experienced innovator or a beginner, there is always room for improvement.

Choose between the short evaluation or the in-depth evaluation and create a baseline to assess corrective actions, find out best practices, and create an action plan. Following the audit or evaluation, an in-person in-depth innovation evaluation, workshops, and other tools can help you further create an action plan to deliver sustainable innovation. The audit, in-person or online, short or in-depth, will have you answer a series of questions and provide you an immediate image and assessment of your strength and potential weaknesses in creating and sustaining innovation.

Various forms of the survey are described: pre- and post-assessment forms suited to longitudinal growth measures and a creativity status assessment form suited for a creativity index. The book guide provides scale and scoring information, and information about how to arrange using the actual survey.

The surveys include traditional scales for originality, creative fluency, and creative flexibility. The survey also presents scales for problem finding, creative self-efficacy, empathy, and collaboration. *Next Generation* refers to the novel and comprehensive qualities of this survey that address the limitations of available creativity surveys and tests.

Pre- and post-forms of the Next Generation Creativity Survey each require about 45 minutes of student time. The Status assessment form also requires about 45 minutes of student time.

EXAMPLE

Example 1: Short Form Innovation Survey*
Example 2: Next Generation Creativity Surveys

CRoC is a research institution dedicated to understanding creative behavior and motivation—and programmatic experiences that can boost creativity. The centers were founded in 2011 by Professors James S. Catterall of University of California and Anne Bamford of the University of the Arts in London, United Kingdom. A featured CRoC project has been the creation and testing of what we call the Next Generation Creativity Survey, an assessment tool that uses traditional self-report scales along with ratings of original student work to assess creative skills and motivations. The survey was piloted during 2012–2013 in eight art, science, and social problem-solving programs

---

* Example 1 is the Short Form Innovation Survey developed by Robert Brands, the innovation coach. Used with permission. See http://www.innovationcoach.com/solutions/ for more details.

across the United States. A second phase of pilot testing is underway this year in 35 art and science programs in Connecticut, New York, Florida, Burbank, Tustin (California), Los Angeles, Wasco (California), Avenal Wasco (California), Kettleman City Wasco (California), Las Vegas, and Nashville. The centers engage in contract program research and evaluation on a limited basis, especially where creative development and motivation are among the program goals. For more details, see http://www.croc-lab.org/.

Quickview Survey Case Studies—Thumbnail Summary of 2014 CRoC Projects

- P.S. ARTS Los Angeles/Central CA Teaching for Creativity
- The Wooden Floor, Santa Ana, California, 10-year longitudinal study
- Musicals schools in New York, Nashville, Las Vegas—creative development
- Big Ideas powered by Disney: Connecticut, New York, Florida, and California—creative development and exploration of teaching for creativity
- Get Lit–Words Ignite, classical and original poetry performance—creativity study
- Inner-City Arts Creativity Laboratory. Research on Teaching for Creativity
- Inner-City Arts/LAUSD/Indiana University U.S. Department of Energy Arts in Education Model Development and Dissemination project. Visual art and theater learning; impacts on academic achievement
- Theater and ASD (autistic children) project, Staten Island, New York. New York Hall of Science Makers Programs.
- 2013–2014 CRoC research involves 35 schools and after-school programs and 1500 children in seven states.

SOFTWARE

Software solutions that are available include

- Development Collaboration Software (New Product Development Software).

- Ideation Software (creating and managing ideas toward the end of the funnel toward actual product development).
- Innovation Culture Software (Corporate Innovation Communication software to create a true sharing and collaborative open environment). For information, contact Software@innovationcoach.com.
- ideaken enables enterprises when they need to collaborate to innovate, with employees, customers, research vendors, academia, or with the global pool of talent. ideaken, a software as a service platform, is powered by innovation-centric collaboration techniques, best practices, and processes bundled into one single service conducive for open innovation. Website: www.ideaken.com.
- Monsoon, an online meeting software designed to help your group overcome the drawbacks of traditional brainstorming.

SUGGESTED ADDITIONAL READING

Brands, R. *Robert's Rules of Innovation*. Hoboken, NJ: Wiley, 2010.
Catterall, J. and Runco, M. *Next Generation Creativity Survey Handbook*. Orlando, FL: Centers for Research on Creativity and The Walt Disney Company, 2010.

26

Scenario Analysis (SA)

Scott Benjamin

CONTENTS

DEFINITION

Scenario analysis (SA) is a process of analyzing possible future events by considering alternative possible outcomes (sometimes called *alternative worlds*). Thus, SA, which is a main method of projections, does not try to show one exact picture of the future. Instead, SA is used as a decision-making tool in the strategic planning process that provides flexibility for long-term outcomes.

USER

This methodology is most effectively used with small groups of 5 to 10 people from different functions within the organization.

OFTEN USED IN THE FOLLOWING PHASES OF THE INNOVATIVE PROCESS

The following are the seven phases of the innovative cycle. An X after the phase name indicates that the tool/methodology is used during that specific phase.

- Creation phase X
- Value proposition phase X
- Resourcing phase
- Documentation phase
- Production phase X
- Sales/delivery phase
- Performance analysis phase

TOOL ACTIVITY BY PHASE

- Creation and value proposition phases—During these phases, SA is used to look at different combinations of designs to determine which alternative provides the best *value add* to the stakeholders.
- Production phase—During this phase, SA evaluates different options of providing the product or service to determine which provides the best overall value for the stakeholders. For example, SA would evaluate if the product should be manufactured within the organization or outsourced.

HOW THE TOOL IS USED

The strategic planning process should be implemented from day one of the innovation process. Any innovator, entrepreneur, top management team, or visionary should be familiar with using the SA tool before investing human or financial capital into an innovation. The innovator should understand what will happen if any variety of situations occur at the industry level before pursuing a specific innovation path. For instance, what will happen to Apple's luxury line of tablets if the economy sees a double-dip recession, remains flat, or rapidly increases? By running through this analysis, the innovator can prepare and guide top management for a variety of organizational outcomes related to developing a new innovative product or implementing a new innovative process.

SA was first made known though Herman Khan's use of it for military purposes during the Second World War on behalf of the RAND Corporation. The tool was developed to determine possible outcomes during military engagement. It was later used as a method of developing various scenarios of possible outcomes of a nuclear war between the United States and the Union of Soviet Socialist Republics (Kahn and Wiener, 1967).

This tool made its way into business decision making around 1970. Pierre Wack of the Royal Dutch/Shell Company began to use the tool as a basis for corporate planning. Shell used the tool to develop various scenarios judging any plausible event that could influence the supply of oil in the coming decades. Shell prepared for the possibility of the development of an oil cartel (Schoemaker, 1995). While at the time this did not seem highly likely, Shell prepared for this scenario. When the Organization of the Petroleum Exporting Countries was formed and began to apply pressure on the industry, Shell was able to quickly change their business plan to implement the changes in the industrial environment (Shell International, 2003).

Since the 1970s, the tool has been used by countless other companies as a method for preparing for and predicting future alternatives. The tool is used in global environmental assessments, political decision making, community engagement, and business planning. This tool has become a source of gaining temporary and sometimes sustainable competitive advantages. Scenario planning allows these companies to become prepared for surprise events or unforeseen occurrences that may change the fundamental assumptions underlying their existing forecasts. This preparation allows for flexible long-term planning and adaptation.

Putting Strategic Planning to Work

SA is a methodology designed to assist strategic planning by creating several potential scenarios for each external factor that may have an impact on organizational direction. Organizations that excel at planning for tomorrow conduct research and development scenarios for many potential future outcomes—it is an integral part of the strategic planning process for any organization.

The purpose of using SA is to improve current decision making by evaluating a series of potential outcomes resulting from different possible scenarios. SA investigates the external forces, events, and competitors that will shape an industry. These variables are then arranged in various permutations in order to simulate a variety of potential future outcomes. Organizations can then use the scenarios in order to form vivid pictures of these outcomes and base strategic decisions on the likelihood and interaction between the various scenarios.

The goal of SA is not to predict the future but to use future predictions to provide a deeper understanding of the variables and circumstances currently in play and those that may come into play that will help inform organizational decision making and direction. Essentially, SA guides long-term planning by developing vivid outcomes through imagining what would happen if certain trends continue and certain conditions are met. This methodology provides preparation for a variety of circumstances and subsequently allows for rapid modifications to a business plan or decision-making process should external factors change in the future.

When, Where, and How to Use It

Organizations frequently develop a business plan or strategic plan as a tool for directing the growth of a company. While these plans are helpful tools, they tend to remain static. In other words, they make the assumption that the external environment will continue to move along in a predictable fashion. Unfortunately, the external environment is made up of various dynamic factors that frequently change and inevitably will have an impact on the original plans of the company. This concept creates a challenge for companies trying to plan without attempting to predict the impact that these future external changes will have on their operation (Porter, 1980). This presents a difficult quandary for innovators developing for tomorrow, and strategic planners and visionaries—How do you create a dynamic strategic plan in an ever-changing external environment?

SA Seven-Step Process

1. Define the scope

Long-term strategic planning is a complex task. The task involves identifying a strategic path for the future and determining what the organization would like to accomplish during a set period of time (Schwartz, 1996). Step 1 of SA involves identifying the goals that the organization is trying to meet and the timeline for accomplishing these goals. SA works most effectively when goals can be relatively specific in nature.

Example

John Q. Innovator was thinking about launching a new business that would provide continuing education classes for medical professionals. He had experience with offering classes in the past as a director of educational services for MedStar, the largest provider of education services in the United States, and believed that he was ready to go out on his own. John sat down at his computer to develop his business plan. His course offerings specialized in additional training for elective plastic surgery. While not a requirement for most physicians, this curriculum would allow physicians to perform additional elective cosmetic surgeries. John saw the migration of online content delivery as an opportunity to become an innovator in the medical education field. There were many decisions to be made with none more important than if this innovative company can compete in the industry. John Q. Innovator decided to utilize scenario analysis in order to assist him in deciding if there indeed is an opportunity in this field, and if the timing for entry was appropriate.

Identify Major Stakeholders

Once the scope of the work has been identified, it is important to assess which stakeholders will have an impact on the goal. As per Edward Freeman, on the *stakeholder theory*, "stakeholder theory argues that there are other parties involved, including governmental bodies, political groups, trade associations, trade unions, communities, financiers, suppliers, employees, and customers. Sometimes even competitors are counted as stakeholders—their status being derived from their capacity to affect the firm and its other stakeholders" (Freeman, 1984). These groups have the ability to influence the future of the organization and, as such, their interests must be included in developing various scenarios.

Example

John Q. Innovator has made a list of the stakeholders that are important to the growth and development of his business.

- Primary shareholder: himself
- Angel investor that has funded his business
- Faculty that he is hiring to teach the medical content
- Doctors that will be his primary customers
- Patients that will be the recipients of the elective surgery
- Regulators that develop the continuing education requirements for medical professionals
- Politicians that influence medical legislation
- Competitors in the industry

Gather Trend Data

The next step is to identify trends that are occurring in the external environment that will affect the future direction of the company. Several strategic management tools have already been developed to assist in conducting external analysis. Referring to macro-environmental tools to help identify additional key stakeholders are the PESTLE analysis and Porter's Five Forces. The PESTLE analysis includes political, economic, sociocultural, technological, legislative, and the environment. Porter's Five Forces direct attention to barriers to entry, the threat of substitution, buyer power, supplier power, and the possibility of substitutable products (Porter, 1980). Using these tools, begin to gather information and develop a series of questions that may have an effect on the success of the venture. Next, briefly explain each trend that will affect the issues at hand. Using this trend, data will help develop a more vivid picture of the current competitive industrial landscape and help form some ideas of threats and opportunities in the industry. Keep in mind that threats may appear on the surface to be a negative consequence of competing but under the surface may present opportunities for future innovation and development.

Example

- How will ObamaCare affect the medical regulations?
- Will the economy respond after the recent recession to allow for additional elective surgeries?
- Will the economy allow disposable income for physicians to attend additional training?

- What is the current legislative environment on medical malpractice lawsuits?
- How have premiums for insurance been affected?
- Are there regulations that make it difficult to become a certified education provider?
- What are the technological changes that will affect the delivery of educational content?
- How has the overall population of surgeons been growing?
- Have the demographics changed to increase or lower the number of high net worth individuals?

Separate Certainties from Uncertainties

At this point, it is important to separate the assumptions that can be validated from those whose outcomes are uncertain. Make a list separating certainties and uncertainties. With the uncertainties, include trends that will significantly affect your business and the factors that may or may not remain constant. For these uncertainties, begin to determine possible outcomes. At this point, do not make these outcomes overly complex. Focus on simplicity. Finally, prioritize these uncertainties and their possible outcome from most important to least important on the list.

Example

Following the recent recession, all economic indicators have shown an increase in disposable income. John Q. Innovator believed that this income should increase the number of elective cosmetic procedures. John also noticed a recent legislative proposal that would allow for the delivery of online content to satisfy the requirements for continuing education. With this eminent change in the regulation, physicians may be more likely to take online education via this more efficient method versus taking time away from their busy practices. However, John also noticed an increase in medical malpractice suits. The ease of filing contingency-based lawsuits against medical practitioners has stimulated the personal injury legal field. How will these uncertainties evolve in the future?

Construct Scenarios

At this point, you should be ready for initial scenario construction. Begin by taking the highest-rated uncertainty from your list and then working your way through the entire list. Develop three possible outcomes for each uncertainty. The first outcome should assume that everything works out for the benefit of your scenario; the second outcome should assume a moderate benefit to the scenario; and the third outcome should assume all things are detrimental to the scenario. It is sometimes helpful to name the various scenarios. For example, when attempting to predict how the economy will affect the growth of the business, you may choose *gold rush, flat,* and *bust cycle.* For instance, this may be helpful as you combine scenarios to see what will happen with a key favorable legislative change in a bust cycle.

Now that you have this matrix developed, determine which scenarios are strategically relevant to the business project and reprioritize your initial list.

Example

ECONOMIC

> *Gold rush*—The economy has shown rapid recovery from 2009 to 2015, and the predictions for a further bull market continue until 2020. Real disposable income measures are predicted to rise consistently at a rate of 5% per year until 2020, and the stock market should continue to grow at double-digit rates. As a result, elective surgeries are expected to continue to increase at a rate of 20% per year through 2020, and the field of plastic surgery will continue to increase in total revenue volume. The growth in the field of plastic surgery will result in an overall increase in new plastic surgeons to the market needing continuing medical education.
>
> *Flat*—The economy has remained flat. Real disposable income levels have remained flat at 2008 levels ($36,000) and are predicted to be at the same level come 2020. Elective cosmetic surgeries have remained flat, and the number of plastic surgeons in 2020 is predicted to be roughly the same as 2015 levels.
>
> *Bust cycle*—The economy has fallen into a double-dip recession. By 2020, disposable income levels are predicted to be $31,000—the lowest level since 1999. As a result, the elective cosmetic surgery segment is predicted to be the lowest level in modern history with many physicians being forced to close up their practices.

Conduct Additional Research

This is the point when many people begin to make decisions using their initial scenarios. This is not recommended. Having identified and classified a number of uncertain situations, it is now proper to dig deeper into the research of each uncertainty in order to identify blind spots and develop more realistic scenarios. Organizations typically have deep knowledge involving areas of expertise related to their core competencies but seldom have the depth of knowledge of peripheral areas to develop vivid scenarios in opportunities that exist on the margins. Sometimes, executing additional research results in pivoting away from the initial innovation and preparing an SA for a related product or service.

Applying Scenarios to Planning

Finally, you combine the various named initial scenarios with your additional research to form a series of scenarios that can be evaluated and tested. By discussing these scenarios with key management team members, you can evaluate both the upside and downside risks of continuing with the innovative process. For this exercise to be successful, the scenarios must resonate with the mental schemas of the key management team members. If they cannot envision these scenarios, the exercise will lose its potency. The scenarios should cover a wide range of possibilities and span to a long-term frame. Various scenarios should be presented to allow for a range of legitimate possibilities and strategic direction.

Example

John Q. Innovator set up a meeting with his primary investor in the medical education venture.

He explained the various scenarios:

With favorable predictions of the economy over the short- to mid-term and the proposed passage of accepting online content delivery for continuing medical education—this scenario showed promise for a successful venture. If the online content delivery method was not passed and the economy proceeded to grow, the competitor analysis showed that there is market share for John to capture using the existing sticks and bricks model of content delivery.

Additional research revealed that opportunities existed in other markets for continuing medical education, including radiology and cardiology. Should the online content delivery market not materialize, John's company can thrive by diversifying into these markets.

EXAMPLES

Insurance: SA continues to play an active role in the insurance industry. The tool is used to increase the organization's risks and capital management capabilities. For instance, scenario testing aids in assessing the simultaneous impact of a set of events such as a global recession and the occurrence of a natural disaster. The tool is used to evaluate long-term solvency of the insurer.

Machine tooling: The machine tools industry is highly cyclical, and developments in a few key end-user markets can have a pronounced effect on sales. Manufacturers of machine tools need a clear understanding of economic developments and the impacts that these developments can have on their investment decisions for future manufacturing runs given a variety of economic conditions. SA helps prepare these tooling shops for cyclical planning through a variety of economic conditions.

StarTrust: SA was used in saving StarTrust. Once completely supported by Enron Federal Credit Union, management at StarTrust conducted an SA to evaluate what would happen if their primary supporter were to collapse. Later, when Enron collapsed, the credit union was saved because managers had taken actions to be less dependent on Enron.

Shell Oil: To build visions of the future, Shell's scenario team focuses on long-term trends in economics, energy supply and demand, geopolitical shifts, and social change. Their work helps Shell understand how consumers, governments, companies, and regulators are likely to behave and respond to developments in the future.

Scenarios play a role in every strategic company decision. In 2005, for example, they raised the probability of a looming gap between society's surging demand for energy and global supplies.

This helped encourage executives to significantly expand production of cleaner-burning natural gas. Shell moved before some other companies to invest in new technology and buy or explore for new gas fields.

AutoNation: SA helped AutoNation survive the 2008 recession. As inexpensive bank financing began to dry up in 2008, the market for automobile sales withered. Many dealers went out of business due to the change in this key industry driver. However, AutoNation experienced profitability and a 400% increase in their stock price. In 2006, the chief executive officer of AutoNation began to run scenarios as to what his business would look like if the financial markets dried up. His analysis included running these low-probability, high-consequence events. As a result, AutoNation was prepared to excel during these periods when his scenarios became reality.

SOFTWARE

- Axiomepm: www.axiomepm.com
- Centage: centage.com
- Decision Path: www.decpath.com

REFERENCES

Freeman, E.R. *Strategic Management: A Stakeholder Approach*. Boston: Pitman, 1984.

Kahn, H. and Wiener, A. *The Year 2000*. New York: Macmillan, 1967.

Porter, M.E. *Competitive Strategy: Techniques for Analyzing Industries and Competitors*. New York: Free Press, 1980.

Schoemaker, P.J.H. Scenario planning: A tool for strategic thinking. *Sloan Management Review*, vol. 36, no. 2, pp. 25–40, 1995.

Schwartz, P. *The Art of the Long View: Planning for the Future in an Uncertain World*. New York: Currency Doubleday, 1996.

Shell International. Scenarios: Explorers' guide. Available at www.shell.com/scenarios, 2003.

SUGGESTED ADDITIONAL READING

Aaker, D.A. *Strategic Market Management*. New York: John Wiley & Sons, 2001. pp. 108 et seq. ISBN 0-471-41572-3.

Chermack, T.J. *Scenario Planning in Organizations: How to Create, Use, and Assess Scenarios*. San Francisco: Berrett-Koehler, 2011.

Day, G.S. and Schoemaker, P.J.H. Scanning the periphery. *Harvard Business Review*, vol. 83, pp. 135–148, 2005.

27

Storyboarding

H. James Harrington

CONTENTS

DEFINITION

Storyboarding physically structures the output into a logical arrangement. The ideas, observations, or solutions may be grouped visually according to shared characteristics, dependencies on one another, or similar means. These groupings show relationships between ideas and provide a starting point for action plans and implementation sequences.

USER

This tool can be used by individuals or groups, but its best use is with a group of four to eight people. Cross-functional teams usually yield the best results from this activity.

OFTEN USED IN THE FOLLOWING PHASES OF THE INNOVATIVE PROCESS

The following are the seven phases of the innovative cycle. An X after the phase name indicates that the tool/methodology is used during that specific phase.

- Creation phase X
- Value proposition phase
- Resourcing phase X
- Documentation phase
- Production phase X
- Sales/delivery phase X
- Performance analysis phase

TOOL ACTIVITY BY PHASE

- Creation phase—During the creation phase, a storyboard is often used to transform an innovative idea into vivid focus. It helps clarify and refine the original concept into a series of pictures that provide significantly more detail than was originally defined in the creative idea. It is used to clarify, detail, and to communicate the original concept.
- Resourcing phase—The storyboard provides a great deal more detail related to the original concept allowing the team to better define the resources that are required to support the original concept.
- Production phase—The storyboard is often used as the blueprint for constructing the product or service. It provides a picture of what the finished output should look like.
- Sales/delivery process—Storyboarding is often used as the outline for a marketing campaign and is sometimes used to provide a potential customer/consumer with an understanding of how the product or service will benefit them.

HOW THE TOOL IS USED

A storyboard is a graphic organizer that provides the viewer with a high-level view of a project or a problem. Storyboards originated in the motion-picture industry to help directors and cinematographers visualize a film's scenes in sequence. Such storyboards resemble cartoon strips. The ideas, observations, or solutions may be grouped visually according to shared characteristics, dependencies on one another, or similar means. These groupings show relationships between ideas and provide a starting point for action plans and implementation sequences.

Storyboarding may be used to give structure to data, ideas, information, observations, etc. Storyboards are used today by industry for planning advertising campaigns such as corporate video production, commercials, a proposal, or other business presentations intended to convince or compel to action. Consulting firms teach the technique to their staff to use during the development of client presentations, frequently employing the *brown paper technique* of taping presentation slides (in sequential versions as changes are made) to a large piece of Kraft paper that can be rolled up for easy transport. The initial storyboard may be as simple as slide titles on Post-it notes, which are then replaced with draft presentation slides as they are created.

General

Storyboarding coordinates *ideas* or *solutions*, thereby demonstrating its application in a problem-solving context. However, it should be noted that storyboarding is an effective technique in many situations, and consider its use in other contexts (Productive Luddite, 2010).

The following are typical types of storyboards:

1. Written descriptive storyboard—A series of separate paragraphs each describing a specific activity that is part of a total series of activities.
2. Stick figures storyboard—A series of separate pictures each showing a specific activity that is part of a total series of activities.
3. Descriptive stick figures storyboard—A series of separate stick figure drawings and the supporting paragraphs related to each of the specific activities that are part of a total series of activities.

4. Landscape storyboard—A series of specific drawings that include the background related to the specific activity and a narrative describing the specific activity that is part of a total series of activities.
5. Photograph storyboard—A series of photographs each related to a specific activity and accompanied by a narrative describing the specific activity that is part of a total series of activities.
6. Boardomatics storyboard—A series of pans and close-up movements on illustrated frames related to a specific activity and accompanied by a narrative describing the specific activity that is part of a total series of activities.
7. Animatics storyboard—A series of two-dimensional art turned into dynamic motion sequences and accompanied by a narrative describing the specific activity that is part of a total series of activities.
8. Cinematics storyboard—A series of three-dimensional animated sequences uniquely suited for high-level testing for spots, viral campaigns, and shorts.

Storyboard types 1, 2, and 3 are the ones most frequently used by the professional innovator. Storyboard type 7 (animatics storyboard) is frequently used during the sales and marketing phases of the innovation process.

Preparation

Record solutions on index cards or another type of notepaper. If solutions have already been developed, transfer each one onto a self-adhesive note or an index card. Otherwise, instruct each member of the group to write his or her ideas onto individual notes or cards. Write only one idea on each self-adhesive note or card. If the group dynamic provides a constructive work environment, you may wish to modify this step by incorporating elements of brainstorming. This could be accomplished by having each participant share ideas with the group as they occur, thereby creating opportunities to build on ideas.

Display the ideas.

1. Post all ideas on a board in random order. The board should be fairly large. Discuss ways to group the notes.
2. Arrange notes in predefined groups.

A list of alternatives is provided below. However, these are only a few suggestions. The group may wish to employ one of them, combine elements of them, or create a scheme of your own. Your decision will be based on the nature of the issue and whether it is best presented in a flow-like format, according to shared characteristics, etc.

Create the Storyboards

- Create the storyboard using either a network, cluster, matrix, or tree structure format.
- Network—Review each idea for similarities or relationships to other ideas. Draw a line between sets of connected or related ideas.
- Cluster—Develop a set of category headings (the number of these varies according to the number of ideas that are being considered; two to six headings is a typical range). The category headings are based on major themes that emerge in the course of developing ideas. Attach the headings across the upper edge of the storyboard. Rearrange the notes or cards by posting them below the appropriate headings. This process will likely generate discussion around ideas that could fall under two or more categories. For resolution of these issues, consider placing the card between two headings, if possible, or apply consensus building or voting rules as appropriate. If a sufficiently large number of ideas fall under multiple categories, it may be more appropriate to use the matrix structure described below.
- Matrix—Develop a list of categories. List all category headings vertically and horizontally. Post ideas according to their fit with the vertically and horizontally listed headings.
- Tree structure—Review ideas and determine which ones must occur before others can be implemented. Arrange cards or notes according to the sequence in which they should occur. The board should flow from left to right. Therefore, begin posting those ideas that must be implemented ahead of others on the left side of the board.

EXAMPLE

See Figure 27.1.

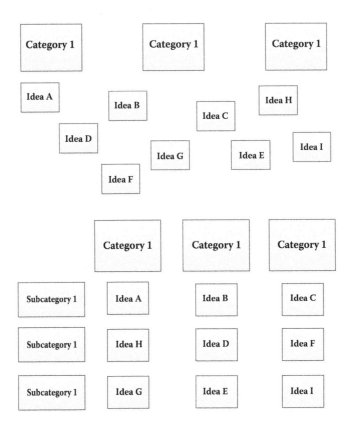

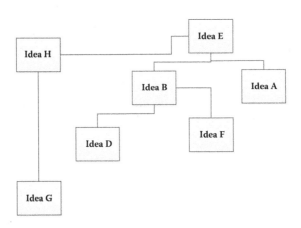

FIGURE 27.1
Ideas breakdown analyses by category.

SOFTWARE

- Atomic Learning's StoryBoard, price: free
- 6sys's Springboard Storyboard, price: $40
- PowerProduction Software's StoryBoard, price: $300
- Forge's Previz Studio, price: between $400 and $900

REFERENCE

Productive Luddite. Storyboard 16:9 Cinema Notebook: Visual Storytelling Technology. Productive Luddite, 2010.

SUGGESTED ADDITIONAL READING

2006 Information Architecture Summit Wrapup. Available at boxesandarrows.com, April 19, 2006.
Cartwright, S. *Pre-production Planning for Video, Film, and Multimedia*. Newton, MA: Hutterworth and Heinemann, 1996.
Cristiano, G. *Storyboard Design Course: Principles, Practice, and Technique*. Hauppauge, NY: Barron's, 2007.

28

Synectics

Frank Voehl

CONTENTS

DEFINITION

Synectics combines a structured approach to creativity with the freewheeling problem-solving approach used in techniques like brainstorming. It is a useful technique when simpler creativity techniques like SCAMPER, brainstorming, and random input have failed to generate useful ideas. It uses many different triggers and stimuli to jolt people out of established mind-sets and into more creative ways of thinking.

USER

This tool can be used by individuals or groups, but its best use is with a group of two to four people. Cross-functional teams can also yield significant results from this activity. Today, Synectics is used by a variety of business people ranging from advertisers and marketers to inventors and designers.

OFTEN USED IN THE FOLLOWING PHASES OF THE INNOVATIVE PROCESS

The following are the seven phases of the innovative cycle. An X after the phase name indicates that the tool/methodology is used during that specific phase.

- Creation phase X
- Value proposition phase X
- Resourcing phase X
- Documentation phase X
- Production phase X
- Sales/delivery phase X
- Performance analysis phase

TOOL ACTIVITY BY PHASE

Synectics is a tool that can be used in all the phases of the innovation cycle when problems occur or when new approaches are being developed. It is particularly useful during the creation phase and the production phase.

HOW TO USE THE TOOL

Synectics is a set of principles and methodologies used to understand and facilitate problem solving and the creative process. The basic idea behind

Synectics is to tap into both the conscious and unconscious creative faculties of the individuals involved, even if they are not creativity professionals. As such, it is a method of identifying and solving problems that depends on creative thinking, the use of analogy, and informal conversation among a small group of individuals with diverse experience and expertise.

Synectics may refer to any kind of group activities involving problem solving and creativity, or may refer to specific techniques promoted under the trademarked name of Synectics. The processes of Synectics evolved from analyzing recordings of business meetings and experimenting with different ways of making the meetings more productive and innovative. Today, Synectics is used by a variety of business people ranging from advertisers and marketers to inventors and designers.

Synectics a useful technique when simpler creativity techniques like SCAMPER, brainstorming, and random input (which are embedded within the Synectics approach) have failed to generate useful ideas, as it uses many different triggers and stimuli to jolt people out of established mind-sets and into more creative ways of thinking.

Putting Synectics to Work

However, given the sheer range of different triggers and thinking approaches used within Synectics, it can take much longer to solve a problem using it than with, say, traditional brainstorming—hence, many experts classify its use as a *second-level tool* when other creativity techniques have failed. The Synectics world process to innovative work is outlined in Figure 28.1.

The problem is that no two experts view this tool in the same way, largely due to the inventor (William Gordon) purposefully incorporating a sense of *vagueness* in order to enhance its flexibility as a creativity-enhancing tool. Generating ideas with Synectics is a three- or four-approach/-stage process:

1. *Referring:* Gathering information and defining the opportunity in terms of direct analogies.
2. *Reflecting:* Using a wide range of techniques to generate ideas, including personal analogies.
3. *Reconstructing:* Bringing ideas back together to create a useful solution using a *compressed conflict* model.
4. *Building fantastic energy:* Users must let their imaginations ramble unrestrained, and connect and concoct the most bizarre solutions imaginable, often described as the *fantastic analogy*.

Synectics© process to innovative work
The intersection of climate, thinking, action

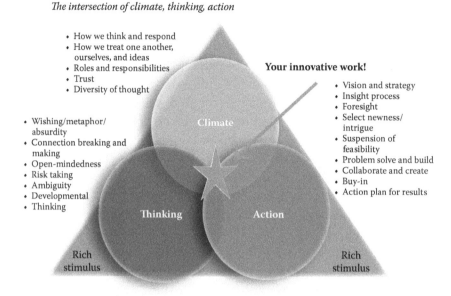

- How we think and respond
- How we treat one another, ourselves, and ideas
- Roles and responsibilities
- Trust
- Diversity of thought

Your innovative work!

- Vision and strategy
- Insight process
- Foresight
- Select newness/intrigue
- Suspension of feasibility
- Problem solve and build
- Collaborate and create
- Buy-in
- Action plan for results

- Wishing/metaphor/absurdity
- Connection breaking and making
- Open-mindedness
- Risk taking
- Ambiguity
- Developmental
- Thinking

Climate

Thinking

Action

Rich
stimulus

Rich
stimulus

FIGURE 28.1
Synectics' role in the innovative process.

In the referring stage, you lay the foundations you will use later for the successful use of the tool. At this stage, you

- Precisely define the problem you want to solve.
- Properly research the factors contributing to the problem.
- Understand what solutions have been tried up to this point.

Reflecting is when you creatively and imaginatively generate possible solutions to the problem you have defined. The emphasis here is on using a range of different *triggers* and *springboards* to generate associations and ideas. Just as with brainstorming, reflecting is best done in a relaxed, spontaneous, and open-minded way, with an emphasis on creative thinking rather than on critical assessment of suggestions. Where Synectics differs from brainstorming and other creativity approaches is in the formal and systematic way it seeks to spark comparison with other approaches and situations, creating new ideas by making associations between these and the problem being solved. That said, a useful way of starting the Synectic idea generation process is to brainstorm inside and around the opportunity or problem normally. This should generate a range of possible solutions to the problem.

If none of these solves the problem, the next step is to use some of the 22 possible triggers described below to try to break free of existing thinking patterns. These triggers reflect things that you can do to transform your current product, service, or approach to try to solve the problem.

Subtract. Simplify. Omit; remove certain parts or elements. Take something away from your subject. Compress it or make it smaller. Think: What can be eliminated, reduced, disposed of? What rules can you break? How can you simplify, abstract, stylize, or abbreviate?

Repeat. Repeat a shape, color, form, image, or idea. Reiterate, echo, restate, or duplicate your reference subject in some way. Think: How can you control the factors of occurrence, repercussion, sequence, and progression?

Combine. Bring things together. Connect, arrange, link, unify, mix, merge, wed, rearrange. Combine ideas, materials, and techniques. Bring together dissimilar things to produce synergistic integrations. Ask: What else can you connect to your subject? What kinds of connections can you make from different sensory modes, frames of reference, or subject disciplines?

Add. Extend, expand, or otherwise develop your reference subject. Augment it, supplement, advance, or annex it. Magnify it: Make it bigger. Think: What else can be added to your idea, image, object, or material?

Transfer. Move your subject into a new situation, environment, or context. Adapt, transpose, relocate, dislocate. Adapt the subject to a new and different frame of reference. Move the subject out of its normal environment; transpose it to a different historical, social, geographical, or political setting or time. Look at it from a different point of view.

Adapt an engineering principle, design quality, or other special quality of your subject to that of another. (The structure of a bird's wing, for example, has served as a model for designing bridges.)

Transfer can also denote transformation. Think: How can your subject be converted, translated, or transfigured? (See also **Metamorphose** and **Hybridize**.)

Empathize. Sympathize. Relate to your subject; put yourself in its *shoes*. If the subject is inorganic or inanimate, think of it as having human qualities. How can you relate to it emotionally or subjectively? Offering helpful insight to an art student, the 18th century German painter Henry Fuseli once advised, "Transpose yourself into your subject."

Animate. Mobilize visual and psychological tensions in a painting or design. Control the pictorial movements and forces in a picture.

Apply factors of repetition, progression, serialization, or narration. Bring life to inanimate subjects by thinking of them as having human qualities.

Superimpose. Overlap, place over, cover, overlay: Superimpose dissimilar images or ideas. Overlay elements to produce new images, ideas, or meanings. Superimpose different elements from different perspectives, disciplines, or time periods on your subject. Combine sensory perceptions (sound/color, etc.).

Think synchronistically: What elements or images from different frames of reference can be combined in a single view? Notice, for example, how Cubist painters superimposed several views of a single object to show many different moments in time simultaneously.

Change scale. Make your subject bigger or smaller. Change proportion, relative size, ratio, dimensions, or normal graduated series.

Substitute. Exchange, switch, or replace: Think: What other idea, image, material, or ingredient can you substitute for all or part of your subject? What alternate or supplementary plan can be employed?

Fragment. Separate, divide, split: Take your subject or idea apart. Dissect it. Chop it up or otherwise disassemble it. What devices can you use to divide it into smaller increments—or to make it appear discontinuous?

Isolate. Separate, set apart, crop, detach: Use only a part of your subject. In composing a picture, use a viewfinder to crop the image or visual field selectively. *Crop* your ideas, too, with a *mental* viewfinder. Think: What element can you detach or focus on?

Distort. Twist your subject out of its true shape, proportion, or meaning. Think: What kind of imagined or actual distortions can you effect? How can you misshape it? Can you make it longer, wider, fatter, narrower? Can you maintain or produce a unique metaphoric and aesthetic quality when you misshape it? Can you melt it, burn it, crush it, spill something on it, bury it, crack it, tear it, or subject it to yet other *tortures*? (Distortion also denotes fictionalizing. See **Prevaricate.**)

Disguise. Camouflage, conceal, deceive, or encrypt: How can you hide, mask, or *implant* your subject into another frame of reference? In nature, for example, chameleons, moths, and certain other species conceal themselves by mimicry: Their figure imitates the ground. How can you apply this to your subject?

Think about subliminal imagery: How can you create a latent image that will communicate subconsciously, below the threshold of conscious awareness?

Contradict. Contradict the subject's original function. Contravene, disaffirm, deny, reverse: Many great works of art are, in fact, visual and intellectual contradictions. They may contain opposite, antipodal, antithetical, or converse elements that are integrated in their aesthetic and structural form. Contradict laws of nature such as gravity, time, etc.

Think: How can you visualize your subject in connection with the reversal of laws of nature, gravity, magnetic fields, growth cycles, proportions; mechanical and human functions, procedures, games, rituals, or social conventions?

Satirical art is based on the observation of social hypocrisy and contradictory behavior. Optical illusions and *flip-flop* designs are equivocal configurations that contradict optical and perceptual harmony. Think: How can you use contradiction or reversal to change your subject?

Parody. Ridicule, mimic, mock, burlesque, or caricature: Make fun of your subject. *Roast* it, lampoon it. Transform it into a visual joke or pun. Exploit the humor factor, Make zany, ludicrous, or comic references. Create a visual oxymoron or conundrum.

Prevaricate. Equivocate. Fictionalize, *bend* the truth, falsify, fantasize. Although telling fibs is not considered acceptable social conduct, it is the stuff that legends and myths are made of. Think: How can you use your subject as a theme to present ersatz information?

Equivocate: Present equivocal information that is subject to two or more interpretations and used to mislead or confuse.

Analogize. Compare. Draw associations: Seek similarities between things that are different. Make comparisons of your subject to elements from different domains, disciplines, and realms of thought. Think: What can I compare my subject to? What logical and illogical associations can I make?

Remember, stretching analogies is a way of generating synergistic effects, new perceptions, and potent metaphors.

Hybridize. Cross-fertilize: Wed your subject with an improbable mate. Think: "What would you get if you crossed a _____ with a _____?"

Creative thinking is a form of *mental hybridization* in that ideas are produced by cross-linking subjects from different realms.

Transfer the hybridization mechanism to the use of color, form, and structure; cross-fertilize organic and inorganic elements, as well as ideas and perceptions. (See also **Metamorphose.**)

Metamorphose. Transform, convert, transmutate: Depict your subject in a state of change. It can be a simple transformation (e.g., an object changing its color) or a more radical change in which the subject changes its configuration.

Think of *cocoon-to-butterfly* types of transformations, aging, structural progressions, as well as radical and surreal metamorphosis such as Jekyll-and-Hyde transmutations.

Mutation is a radical hereditary change brought about by a change in chromosome relations or a biochemical change in the codons that make up genes. How can you apply metamorphosis or mutation to your subject?

Symbolize. How can your subject be imbued with symbolic qualities?

A visual symbol is a graphic device that stands for something other than what it is. (For example, a red cross stands for first aid, a striped pole for a barber shop, a dove bearing an olive branch for peace, etc.)

Public symbols are clichés insofar as they are well known and widely understood, while private symbols are cryptic and have special meaning only to their originator. Works of art are often integrations of both public and private symbols.

Think: What can you do to turn your subject into a symbolic image? What can you do to make it a public symbol? A private metaphor?

Mythologize. Build a myth around your subject.

In the 1960s, pop artists *mythologized* common objects. The Coca-Cola bottle, Brillo pads, comic-strip characters, movie stars, mass-media images, hot rods, hamburgers and French fries, and other such frivolous subjects became the visual icons of 20th century art.

Think: How can you transform your subject into an iconic object?

Fantasize. Fantasize your subject. Use it to trigger surreal, preposterous, outlandish, outrageous, bizarre thoughts. Topple mental and sensory expectations. How far out can you extend your imagination?

Think *what-if* thoughts: What if automobiles were made of brick? What if alligators played pool? What if insects grew larger than humans? What if night and day occurred simultaneously?

Creative transformation demands an iconoclastic attitude. To invent, one must be contrary and go against established conventions and stereotypes. Remember, inventors create great inventions only by breaking the *rules*.

Art-Think: Ways of Working

1. Identity: Set the problem or task; identify the subject.
2. Analyze: Examine the subject, break it down, classify it.
3. Ideate: Think, fantasize, produce ideas. Generate options toward a creative solution. Relate, rearrange, reconstruct.
4. Select: Choose your best option.
5. Implement: Put your ideas into action. Realize it. Transform imagination and fantasy into tangible form.
6. Evaluate: Judge the result. Think about new options and possibilities that have emerged. Go back to step 1.

Nicholas Roukes
Design Synectics: Stimulating Creativity in Design

Synectics Viewed as a Diamond

The Synectics tool example shown in Figure 28.2 was developed by Synecticsworld and is in the form of a process diamond. It begins with springboards for laying out the task, exploring the options available, selecting the best alternatives, developing the details, and possible solutions.

SUMMARY

The term *Synectics* is derived from a Greek word that roughly means "bringing together disparate ideas that may appear unrelated." Synectics aims to understand and facilitate the cooperation of different people with differing ideas in order to develop creative solutions to problems.

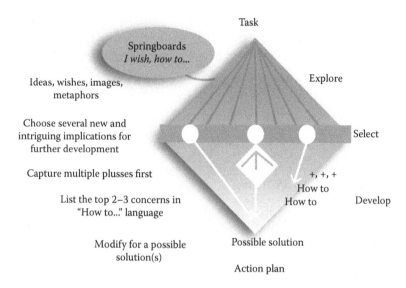

FIGURE 28.2
Synectics viewed as a diamond.

The study and application of Synectics is based on three assumptions:

1. The creative process is not relegated to innate talent or special ability. Instead, it is something that can be deliberately taught.
2. The creative process is the same, whether it is applied to works of art or scientific inquiry.
3. The creative process works the same way for both individual and group endeavors.

The study of Synectics can provide insight into the way creativity happens, as well as techniques to facilitate creative problem solving. The world of Synectics continues to evolve as scholars and business people apply its principles in innovative new ways. Use triggers as starting points for brainstorming. Again, once you have done this, evaluate whether you have a satisfactory solution to the problem you are addressing. If you have not, it is time to move to the next stage: using *Synectic springboards* to stimulate new ideas. These are analogies between the current situation and other situations or things.

They can be functional analogies (with other products, services, and approaches that do a similar job), analogies with other phenomena (e.g., with an ocean storm, a rainforest, or a mechanical digger), or stretched analogies (e.g., comparisons with emotions or symbols). Reconstructing is where you collect all of the ideas you have created during the *reflecting* step, and evaluate them rationally, bringing them together to create practical and useful ideas.

EXAMPLE

Synectics Case Study: Unilever Laundry Products

Background: According to Synecticsworld, the tool worked with the Unilever team to not only help the company reinvent their laundry products category, but also to build an innovation funnel that Unilever employees throughout the world could connect with to champion ideas.

Process: The Synecticsworld-led Unilever team, made up of 42 representatives from 11 countries, worked for 2.5 days.

On day 1, they learned the Synectics tool/techniques and began generating initial ideas, explored consumer segments, and exposed some of their individual, consumer-related stereotypes.

On day 2, the main group broke up into eight smaller groups and visited the homes of consumers, while the second team of experts floated between the locations providing facilitation and coaching where necessary. The group combined the insights they had just developed with the starting ideas from the previous day. The afternoon was spent on idea development and using creativity as a means of enriching the concepts.

Day 3 brought new consumers into the process and used their insights and perspectives to build the existing ideas into even better ones. The groups had an opportunity to take the feedback from consumers and work up the concepts once more before sharing the ideas in a marketplace with the entire group.

What emerged was a highly energetic, charged group who had collaboratively developed an innovation funnel of new and exciting insight-based ideas, and were prepared to drive selected concepts in their individual markets.

Results: After the workshops, the teams presented their work at two separate meetings, one with Unilever marketing managers and directors, and the other with laundry category executives. "Both meetings went very well and our approach to innovation and the ideas themselves were enthusiastically endorsed," affirmed Toffael Rashid, Unilever.

SOFTWARE

- *The Creative Thinker by Idon Resources.* This Synectics-type software brings you nonlinear, yet structured, thinking with the ability to further develop your thoughts in the form of user-friendly and fun to use graphical hexagon modeling. You can even organize your knowledge, supplement ideas with notes, and directly and meaningfully link ideas to the Internet, documents, slide presentations, spreadsheets, and more. The Creative Thinker literally allows you to visualize your thoughts, rapidly access knowledge, and combine the two creatively in real time for insights (see http://www.idon resources.com/ct/creativethinker.html).

- *Creative thinking program.* This program can be customized to be anywhere from two days to six months. The objective is to use Synectics to drive innovation, improvisation, and creative thinking into a team or organization. A combination of skill development training, real-time process facilitation, team and individual coaching, and interviews. This program includes a repeatable creative thinking process that can be used by the different teams, along with relative thinking best practices, tools, and techniques for individuals and teams (http://www.creativeemergence.com/creativethinking.html).

- *X-Mind.* You can use simple mind maps if you choose, or *fishbone*-style flowcharts if you prefer. You can even add images and icons to differentiate parts of a project or specific ideas, add links and multimedia to each item, and more (see http://www.xmind.net/).

REFERENCES

Gordon, W.J.J. *Synectics: The Development of Creative Capacity,* 1960.

Roukes, N. *Design Synectics: Stimulating Creativity in Design.* Worcester, MA: Davis Publications, 1988.

SUGGESTED ADDITIONAL READING

Mauzy, J. and Harriman, R. *Creativity Inc.: Building an Inventive Organization*. Boston: Harvard Business School, 2003.

Nolan, V. *The Innovators Handbook*. New York: Penguin, 1989.

Nolan, V. and Williams, C. *Imagine That!* New York: Publishers Graphics, LLC, 2010.

Prince, G.M. *The Practice of Creativity*. New York: MacMillan, 1970.

Prince, G.M. *The Practice of Creativity: A Manual for Dynamic Group Problem-Solving*. Battleboro, VT: Echo Point Books & Media, LLC, 2012. 0-9638-7848-4.

29

TRIZ

H. James Harrington

CONTENTS

DEFINITION

TRIZ (pronounced *treesz*) is a Russian acronym for *teoriya resheniya izobretatelskikh zadatch*, the theory of inventive problem solving, originated by Genrich Altshuller in 1946. It is a broad title representing methodologies, tool sets, knowledge bases, and model-based technologies for generating innovative ideas and solutions. It aims to create an algorithm to the innovation of new systems and the refinement of existing systems, and is based on the study of patents of technological evolution of systems, scientific theory, organizations, and works of art.

USER

This tool can be used by individuals or groups, but its best use is with a group of four to eight people. It originally was applied primarily to new product design, product evolution improvement, and technical problem solving; as a result, it was primarily used in an engineering environment. During recent years, it has been expanded to cover processes, and systems improvement and development expanding its usage to cross-functional teams.

OFTEN USED IN THE FOLLOWING PHASES OF THE INNOVATIVE PROCESS

The following are the seven phases of the innovative cycle. An X after the phase name indicates that the tool/methodology is used during that specific phase.

- Creation phase X
- Value proposition phase
- Financing phase
- Documentation phase
- Production phase X
- Sales/delivery phase
- Performance analysis phase

TOOL ACTIVITY BY PHASE

- Creation phase—TRIZ is most effectively used as a tool to help design new products or to improve existing products or services.

- Production phase—During the production phase, TRIZ is frequently used to solve manufacturing problems and to improve manufacturing procedures.

HOW TO USE THE TOOL

TRIZ: Benefits and Outcomes

Table 29.1 lists the 14 benefits of TRIZ, shown as outcomes, based on the writings of Altshuller as adapted by Isak Bukhman (Bukhman, 2012).

Overview of TRIZ

During the 1940s, a Russian engineer named Genrich Altshuller directed his research to understand how engineered systems have evolved over time. After reviewing more than 40,000 patents, the result of this study was a theoretical-based approach known as *patterns of evolution of technological systems*. In the early 1980s, Genrich Altshuller researches shifted away from technology and started focusing on areas of creativity. This brought about the end of what was known as the *classical TRIZ* era.

One of the conclusions that Altshuller came up with was that problems for which there were no known means of solving them (inventive problems) involved one or more *contradictions*. He believes that if an engineer could define the contradictions and solve one or more of them, the system will advance to the next step in its evolutionary course.

This led to him to develop the *laws of system evolution*, which are the foundations of TRIZ. There are three problem-solving groups involved:

1. Inventive and separation principles
2. System of standard solutions
3. Algorithm for inventive problem solving (ARIZ 85-C)

ARIZ is a methodology used to analyze a problem directed at reviewing, formulating, and solving contradictions. As ARIZ was worked on and refined, the algorithm grew to more than 60 steps. A high-level generic flowchart of ARIZ to solve a problem with minimal change to a system is shown in Figure 29.1. The specific steps of ARIZ with an example are shown in the Example section of this chapter.

TABLE 29.1

TRIZ Benefits and Outcomes

Benefits and Outcomes	Structural Component
1. TRIZ is a science of system development based on laws of systems evolution and the best practices of thousands of developers and scientists.	Laws of system evolution
2. TRIZ helps realize the privilege and obligation each member of our society has to be a creative person and to live a creative life. At some point in the past, someone did something that makes our lives today more comfortable.	Creative person development
3. TRIZ allows us, in turn, to do something useful for the world now and for generations to come.	Scientific effects and patents
4. TRIZ increases the speed of system development and evolution. This is a primary global function of TRIZ because technological evolution reflects and propels the development of our civilization.	Techniques of breaking psychological inertia
5. TRIZ helps model problems that are not well defined into a specific problem that can be solved by any engineer.	System of standard solutions
6. TRIZ is a natural amplifier of our talents, knowledge, and experience. Everything that we do in our life and any decision that we make will be better and more effective when we use TRIZ.	Creative person development
7. TRIZ changes the people that learn and use it. They become more inventive and creative.	Creative person development
8. TRIZ has no limitations (except one). It can be applied to any new or existing system undergoing development.	Inventive and separation principles
9. TRIZ has only one limitation … the limitations of the physical world. Even in this situation, TRIZ can help find way to overcome scientific limitations.	Techniques of breaking psychological inertia
10. TRIZ breaks psychological inertia, the main innovation killer.	Techniques of breaking psychological inertia
11. TRIZ technology for innovation is the process of using all parts of TRIZ in combination with other proven methods and best practices of effective project teams for system development and problem solving.	Inventive and separation principles
12. TRIZ technology for innovation is applying through TRIZ innovation roadmaps for project creation and problem solving.	Algorithm for inventive problem solving (ARIZ 85-C)

(Continued)

TABLE 29.1 (CONTINUED)

TRIZ Benefits and Outcomes

Benefits and Outcomes	Structural Component
13. TRIZ innovation roadmap for project creation and problem solving is a combination of parts of TRIZ, along with other proven methods and best practices of effective project teams. When applied in the most effective sequence, it will lead to the achievement of best results for any given project or problem.	Algorithm for inventive problem solving (ARIZ 85-C)
14. A TRIZ innovation roadmap is a complete set of tools for the conceptual stage of product/process/service design.	Algorithm for inventive problem solving (ARIZ 85-C)

ARIZ: The TRIZ Algorithm

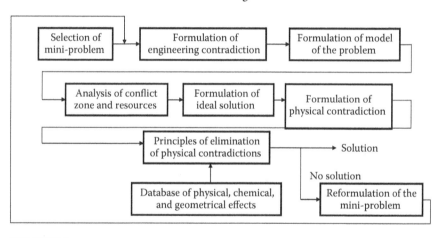

FIGURE 29.1
TRIZ model.

Altshuller's initial focus was on technical contradictions. He identified 39 system characteristics associated with technical contradictions. As a result, more than 1250 technical contradictions were defined for which the traditional resolution was compromise. To simplify this complexity, he developed the 40 Inventive Principles (see Figures 29.5 and 29.10). These principles are all directed at defining potential ways for changing the system in order to eliminate a targeted contradiction. To simplify the conditions/options that are available to correct contradictions, he established a contradiction table that helps the individual using the methodology to reduce the number of options that are available based on the particular type of contradiction he or she is working with.

When Altshuller directed his focus to physical contradictions, he identified another set of principles called *separation principles*.

In the mid-1970s, Altshuller developed a tool for modeling problems called *substance-field analysis* (also called Su–field analysis). He concluded a properly functioning system consisting of substances and the energy that causes substances to act on each other (fields). This led to the establishment of a set of 76 Standard Solutions that are most frequently used to solve problems of this nature. Because the 76 Standard Solutions are not related to specific technologies through the use of analytic thinking, they can nobly be transferred to effectively define solutions between different branches of technology.

To provide additional technical direction, he developed several comprehensive collections of *technological effects* and *phenomena*. These were based on technical principles and effects that he recognized as proving to be useful in solving problems and creating new products over the years.

This may sound complex and difficult to understand, and it is at first. Altshuller's creative mandate and original thinking has resulted in some new definitions and thought patterns that most people have not been acquainted with. But the effort and time required to familiarize yourself with these new concepts is well worthwhile. Yes, it is not as simple and may take a little longer than brainstorming to solve a problem, but by using these approaches most groups come out with far better and effective solutions. In cases when a reasonable solution is not definable by common and normally used innovative approaches, the TRIZ methodology will solve these seemingly unsolvable situations.

Basically, there have been three distinct evolutions in the TRIZ methodology during the last 70 years:

- Classical TRIZ—Development led by Altshuller from the mid-1940s to the mid-1980s.
- Contemporary TRIZ phase I—Developed during perestroika in the former Soviet Union from the middle of the 1980s to the early 1990s.
- Contemporary TRIZ phase II—Penetration into the Western world from the early 1990s to the present.

To minimize some of the confusion, I present the following definitions that will be used throughout the remainder of this chapter.

KEY TRIZ DEFINITIONS AND INFORMATION

40 TRIZ principles—These are 40 statements that describe approaches to resolving technical conflict (problems or contradictions) that were defined by Genrich Altshuller on the basis of his study of more than 200,000 patents (a list of these principles with examples is shown in Table I of the Example section of this chapter). These 40 TRIZ principles have a twofold purpose:

- Within each principle resides guidance on how to conceptually or actually change a specific situation or system in order to be rid of a problem.
- The 40 principles also train users in analogical thinking, which is to see the principles as a set of patterns of inventions or operators applicable to all fields of study.

39 Characteristics of a technical system—These are the 39 system characteristics for expressing technical contradictions defined in the late 1960s.

Patterns of evolution—These are sets of terms or statements that define trends that have strong, historically recurring tendencies in the development/evolution of man-made systems. A beginner's guide on how to use these patterns is shown in the Example section of this chapter.

Lines of evolution—These describe in greater detail the typical sequences of stages (positions on a line) that a system follows, a specific pattern of evolution in the process of its natural progress. Once these positions are known, the system's current position on a line can be identified, and the possibility of transitioning to the next positions can be assessed, for example, become flexible or use micro-level properties of materials utilized. Lines of evolution are grouped under one or more of the patterns of evolution that they support.

Ideality—Ideality is the ratio between useful and harmful effects. The general direction of system improvement maximizes the ratio of ideality. We improve the level of ideality as useful inventive solutions are created and implemented. You increase ideality by increasing the number or magnitude of useful functions or by reducing the number or magnitude of harmful functions. This technique is only used with I-TRIZ.

Altshuller's 40 TRIZ Principles and His 39 Engineering Parameters (Altshuller, 1997)

Because the 40 TRIZ principles and the 39 characteristics of a technical system are key elements in defining solution to problems and evaluating opportunities, a summary of each is listed in Figures 29.2 and 29.3, respectively.

How to Use I-TRIZ

There typically are nine parts or steps in the process of solving a problem using Altshuller's approaches (Bukhman, 2012):

- Part 1. Analyzing the problem
- Part 2. Analyzing the problem model
- Part 3. Formulating the ideal ultimate results and physical contradictions
- Part 4. Mobilizing the utilizing substance—field resources
- Part 5. Applying the knowledge base
- Part 6. Changing or substituting the problem
- Part 7. Analyzing the model for resolving the physical contradictions
- Part 8. Capitalizing on the solution concept
- Part 9. Analyzing the problem-solving process

Altshuller's 40 TRIZ principles for conflict resolution	
1 Segmentation	21 Rushing through
2 Extraction	22 Convert a harm into a benefit
3 Local conditions	23 Feedback
4 Asymmetry	24 Mediator
5 Combining	25 Self-service
6 Universality	26 Copying
7 Nesting	27 Disposable object
8 Anti-weight	28 Replacement of a mechanical system
9 Prior counteraction	29 Use a pneumatic or hydraulic construction
10 Prior action	30 Flexible film or thin membranes
11 Cushion in advance	31 Use of porous material
12 Equipotentiality	32 Changing the color
13 Inversion	33 Homogeneity
14 Spheroidality	34 Rejecting and regenerating parts
15 Dynamicity	35 Transformation of physical and chemical states
16 Partial-excessive action	36 Phase transition
17 Shift to a new dimension	37 Thermal expansion
18 Mechanical vibration	38 Use strong oxidizers
19 Periodic action	39 Inert environment
20 Continuity of a useful action	40 Composite materials

FIGURE 29.2
Altshuller's 40 TRIZ principles.

39 Engineering parameters for expressing technical contradictions	
1 Weight of moving object	21 Power
2 Weight of nonmoving object	22 Waste of energy
3 Length of moving object	23 Waste of substance
4 Length of nonmoving object	24 Loss of information
5 Area of moving object	25 Waste of time
6 Area of nonmoving object	26 Amount of substance
7 Volume of moving object	27 Reliability
8 Volume of nonmoving object	28 Accuracy of measurement
9 Speed	29 Accuracy of manufacturing
10 Force	30 Harmful factors acting on object
11 Tension, pressure	31 Harmful side effects
12 Shape	32 Manufacturability
13 Stability of object	33 Convenience of use
14 Strength	34 Repairability
15 Durability of moving object	35 Adaptability
16 Durability of nonmoving object	36 Complexity of device
17 Temperature	37 Complexity of control
18 Brightness	38 Level of automation
19 Energy spent by moving object	39 Productivity
20 Energy spent by nonmoving object	

FIGURE 29.3
Thirty-nine characteristics of a technical system.

Part 1. Analyzing the problem. The object of part 1 is to get a clear, precise understanding of all the parameters related to the problem/opportunity. The main purpose of part 1 is to facilitate the transformation from a vague problem situation to an efficient and useful model—the problem model.

- Formulate the mini-problem (consider solving the problem with minimal change to the system).
- Define the functional conflict in elements—Conflict in elements could include a workplace, or tools.
- Geographic model for further analysis.
- Exaggerate the conflict by indicating the limiting states (action) of elements.
- Use Su–field analysis for the problem model to include the conflicting elements and exaggerated formula of conflict, what should be rendered by an additional element, reference as to the X element, to be introduced into the system to solve the problem (e.g., what the exit element should hold, preserve, eliminate, improve, provide, etc.).
- Apply the system of standard solutions—Consider solving the problem model using the system of standard solutions. If this does not solve the problem, go to part 2.

Part 2. Analyzing the problem model. The main purpose of part 2 is to create an inventory of available resources (space, time, substances, and feels) that may be applied toward solving the problem.

- Define the operational zone (OZ)—In the simplest case, the OZ is the space where the conflict indicated in the problem model appears.
- Define the operational time—Operational time is the available *time resource*, consisting of the time during which the conflict occurs and the time before and after the conflict.
- Define the substance and field resources—Define the substance and field resources (SFR) of the analyzed system, the environment, and the workplace. Create a list of resources.

Part 3. Formulating the ideal ultimate results and physical contradictions. Working through part 3 should result in the development of an image of the ideal ultimate result (IUR).

- Identify IUR-1—Identify and document IUR-1. With the exception of the conflict, harmful action is associated with the useful one. Other types of conflict are possible; thus, attaining a new useful feature or eliminating harmful one should not be compromised by either the degradation of other features or the appearance of a new harmful feature.
- Exaggerate IUR-1—For a mini-problem, the introduction of new substances and fields in the system is prohibited; only the SFR should be utilized.
- Formulate the physical condition for the micro-level—If the problem cannot be solved at the micro-level, there may be no way to formulate the activity.
- Formulate the IUR-2.

Part 4. Mobilizing the utilizing substance—field resources. Part 4 consists of systematic procedures directed toward increasing the availability of resources by providing derived SFR that can be obtained almost free of charge through combinations of the resources that are available.

- Simulate with smart little creatures—Create a graphic model of a conflict using smart little creatures (stick-figure drawings). Modify the graphic model so that the smart little creatures act without conflict.
- *Stepping back* from the IUR—If you know what the desired systems should be and the only problem is finding a way to obtain this system, it is often helpful to *step back* from the IUR.

- Using a mixture of substance results—It may now be possible to solve the problem using a mixture of substance results.
- Using empty space—Try solving the problem by substituting resources with either empty space or a mixture of substance resources and empty space.
- Using an electrical field—Determine if the problem can be solved by introducing an electrical field or to interacting electrical field rather than substances.
- Using a field and field-sensitive substance—Try solving the problem using a field and a substance or a substance added that is responsive to this field.

Part 5. Applying the knowledge base. The purpose of part 5 is to mobilize all experiences associated with the TRIZ knowledge base. The problem at hand is significantly clearer at this point, so direct utilization of the knowledge base may be successful.

- Consider solving the physical problem by applying the system of standard solutions.
- Consider the problem by applying solution concepts to nonstandard problems that have already been solved using ARIZ.
- Consider resolving the physical contradictions by using the separation principle.
- Consider resolving the physical contradictions by utilizing the library of effects and phenomena.

Part 6. Changing or separating the problem. Often simple problems can be solved by actually overcoming the physical contradiction, for example, by separating contradictory requirements in time or in space.

- If the problem is solved, transform the theoretical solution concept into a practical one: formulate the principle of action and create a schematic diagram of a device that implements this principle.
- If the problem is not solved, check to see whether the description in part 1 represents a combination of several problems.
- If the problem is still not solved, select another technical contradiction.
- If the problem is still not solved, return to part 1 and formulate the mini-problem with respect to the supersystem or subsystem.

Part 7. Analyzing the method for resolving the physical contradiction. During part 7, the quality of the selected solution(s) will be evaluated. Since existing resources are better utilized, the solution should be cost-effective. If it is not, it is better to spend a few more hours to obtain a new more practical solution.

- Check the concept of the solution—Review each introduced substance and field to be sure they are necessary. Is it possible to apply available or disrupted SFR instead of introducing new and additional substances or fields?
- Preliminary estimate of the solution—Review the proposed solution considering the following:
  - What physical constraints were resolved by the proposed solution?
  - How well will the proposed solution be accepted by the affected people?
  - Have the requirements of IUR-1 been satisfied?
  - Does the solution consist of at least one control element? If so which element? How is it controlled?

If you cannot answer positively to these questions, a new solution needs to be developed.

- Check the novelty of the proposed solution versus a patent search—To protect the knowledge capital of the organization, unique items could start a patent process.
- Determine what subproblems might appear during the development of the new technological system—Subproblems often require major changes to the proposed solution to overcome challenges associated with implementation.

Part 8. Capitalizing on the proposed solution. Often, an innovative idea not only corrects the original problem but it also provides a universal key to many other problems. The purpose of part 8 is to maximize utilization of resources by sharing the proposed solution with other potential users.

- Find how supersystems that encompass the changed system should be changed.
- Check whether the changed system or supersystem can be applied to a new application.
- Apply the proposed solution concept to solve other problems—Develop a general solution principle and consider direct application of the solution principle to other problems. Create a matrix that includes all possible modifications of the solution concept, and consider every combination produced by the matrix.

Part 9. Analyzing the problem-solving process. With every process, a postmortem analysis should be conducted to capture the experience learned so that future cycles through similar processes will be enhanced. The purpose of part 9 is to conduct this analysis, recording positive and negative experiences related to the nine parts of the TRIZ process.

- Compare the process that was just executed and compare it to the theoretical perfect process. Note all deviations, both positive and negative.
- Compare the solution concept that was developed to the TRIZ knowledge base. If the knowledge base does not include the presently used principal, add the principal to the knowledge base.

Summary of the Nine Parts

These nine parts provide a systematic way to change the thought patterns of the people using it to solve both simple and complex problems. It has been used literally thousands of times and often has solved problems that were classified as unsolvable. It is a complex process but one that is well worth using. There now are a number of software programs that greatly simplify the process while taking advantage of information that has been accumulated during the past 70 years.

SYSTEM (TECHNICAL) CONTRADICTION MATRIX

This matrix uses the 39 engineering parameters for expressing technical constraints and the 40 TRIZ principles. Both the vertical and the horizontal lines on the matrix are made up of the 39 engineering parameters for expressing technical contradictions. Vertical terms are used to pick an action that you want to consider using to improve the item you are working on. The horizontal axis is used to identify negative things that could occur based on the proposed solution the team is looking at. At the junction of the horizontal and vertical columns, up to four TRIZ principles are listed that are the most likely to offset the contradiction based on past experience.

You can readily understand the complexity of the system contradiction matrix as it could result in more than 60,000 different combinations

(39 × 39 × 40). Limiting the number of principles from zero to four at each junction point of the 39 × 39 greatly reduces the complexity and simplifies the problem-solving effort the team is subjected to when using the matrix. When you consider the size and complexity of the matrix, it would have to be reduced down so small that is not readable on a page in this book. It is typically printed out on a sheet of paper that is 20 × 20 and with the larger paper is very readable. Included is a foldout that will give you a readable view of the system contradiction matrix.

Do not let the matrix overwhelm you. It is really quite easy to use once you understand the matrix. It is as simple as A, B, and C.

- Step A. Define the characteristic that you will be using.
- Step B. Define the worsening characteristics that are targeted for evaluation.

  With the matrix, determine the crossover point between A and B (point C), the intersecting point where one to four basic principles are listed that should be considered to offset the worsening parameter.
- Step C. At the crossover, 4 of the 40 principles that most likely would offset the impact of contradictions upon the system are listed.

  Analyze each of the recommended 40 principles to determine how each one would influence the contradiction. Although all four TRIZ principles are seldom used to offset the worsening parameter, often one or two of them leads to the right answer.

To help you understand how to use it, we have taken only the first 10 engineering parameters for expressing technical constraints and plotted one of them on the horizontal axis and the other on the vertical action description (see Figure 29.4). The team has selected the seventh engineering parameter for expressing technical constraints (#7—volume of a moving object) as a potential activity to take advantage of the improvement opportunity (point A). Horizontally at the top of the matrix, 10 of the 39 engineering parameters for expressing technical constraints are listed. These are characteristics that could get worse as a result of the proposed improvement activity. For this example, the team selected point B (#4—length of a stationary object). Horizontal column "A" intersects with vertical column "B" at point "C." At this intersection, the contradiction matrix recommends that you evaluate the following four principles to eliminate the contradiction or at a minimum minimize the impact the contradiction

	Weight of a mobile object	Weight of a stationary object	Length of a mobile object	Length of a stationary object (B)	Area of a mobile object	Area of a stationary object	Volume of a moving object	Volume of a stationary object	Speed	Force
	1	2	3	4	5	6	7	8	9	10
1. Weight of a mobile object										
2. Weight of a stationary object										
3. Length of a mobile object										
4. Length of a stationary object										
5. Area of a mobile object										
6. Area of a stationary object										
7. Volume of a moving object A				C						
8. Volume of a stationary object										
9. Speed										
10. Force										

FIGURE 29.4
Condensed view of system (technical) contradiction constraints.

has on the system. In this example, the four principles that are recommended by the contradiction matrix are as follows:

1. 2 = Weight of a stationary object
2. 8 = Volume of a stationary object
3. 14 = Strengths
4. 35 = Adaptability

Two Common Approaches to Using the TRIZ Matrix

There are two common approaches to using the TRIZ matrix. They are

1. Worsening parameter analysis
   - Define the improvement parameter that is considered for implementation in order to define the parameters that could result in having a negative impact on the system.
   - Review the 39 (engineering parameters for expressing technical constraints) potential worsening parameters to define the specific parameters that could relate to the individual initiative that you are working on.
   - Look at the intersection between the identified worsening parameters and the proposed improvement parameter to identify TRIZ principles that could be used to offset the worsening parameter.
   - Then evaluate each of the recommended TRIZ principles to determine if it can be used to minimize or eliminate the impact of the worsening parameter.
   - Repeat this process for each of the identified potential worsening parameters.
2. Analysis of combined worsening parameters
   - Define the improvement parameter that you are considering using.
   - Analyze all of the recommended TRIZ principles that are in the horizontal line related to the improvement principle that is being evaluated.

Define how many times each day individual TRIZ principles are recommended for analysis across the entire horizontal line. On the basis of this analysis, create a list of how many times a specific TRIZ principle is recommended as a potential solution for the negative parameters.

Usually by applying one or two of these principles, the team is able to offset the negative impact that the contradiction will have on the system. In this example, we looked at only one of the characteristics that could get worse from changing *the volume of a moving object*. There are more than 30 other characteristics defined in the matrix that could get worse if the volume of a moving object is changed. As a result, it is necessary to look at all of the characteristics that could get worse as a result of the action the team is taking. Typically, the characteristics that have the highest possibility of occurring are the only ones for which the team will perform a related detailed analysis. My personal approach to this situation is to take the proposed improvement action and make a list of all the principles that are recommended at the 39 junctions, with the characteristics that could get worse. I then sum this up to determine which of the 40 TRIZ principles are most often recommended for correcting the contradiction. Typically, I would analyze two to six of the most frequently recommended TRIZ principles.

How TRIZ Evolves

TRIZ has continuously evolved since Alzheimer's developed the original concepts. The following is a list of tools that most of TRIZ professionals use in place of some of the basic tables.

Trends (laws) and subtrends (lines) of technological system evolution

- Trend of increasing degree of ideality
- Trend of non-uniform evolution of subsystems
- Trend of completeness of system parts
- Trend of *energy conductivity* of systems
- Trend of harmonization of rhythms
- Trend of transition to supersystems
- Subtrend of transition from mono- to bi- and poly-systems
- Subtrend of increasing structurization of voids
- Trend of increasing dynamism
- Lines of increasing dynamism
- Trend of increasing substance–field interactions
- Lines of evolution of su–fields
- Trend of transition from macro- to micro-levels
- Trend of matching–mismatching (coordination–non-coordination)
- General pattern of engineering systems evolution

This is part of the TRIZ Body of Knowledge that was prepared by Simon Litvin, Vladimir Petrov, Mikhail Rubin, and Victor Fey. For more detailed information related to these laws and tools, contact the Altshuller Institute at the following:

- URL: http://www.aitriz.org/
- By phone (508-799-6601)
- Or contact Victor Fey at fey@trizgroup.com

How to Use the Contradiction Table

Step 1. From the row entitled "Feature to Improve," find a relevant characteristic.

Step 2. Locate a column that corresponds to the "Undesired results," the characteristic that degrades as a result of improving the above feature.

Step 3. Refer to the 40 Inventive Principles whose numbers are listed in the intersecting cell.

EXAMPLES

To provide you with a better understanding of the 40 Inventive Principles, we will take a detailed look at a very small sample of them.

Inventive Principle #16—Partial or excessive action. If it is difficult to obtain 100% of a desired effect, achieve somewhat more or less than this in order to greatly simplify the problem.

> Example 1: A cylinder is painted by being dropped into paint but consumes more paint than desirable. Excessive paint is then removed by rapidly rotating the cylinder.
>
> Example 2: To obtain uniform discharge of a metallic powder, the hopper has a special internal funnel that is constantly overfilled to provide nearly consistent pressure.

Inventive Principle #19. Periodic action

- Replace a continuous action with a periodic (pulse) one.
- If an action is already periodic, change its frequency.
- Use pulses between impulses to provide additional action.

Example 1: An impact wrench loosens corroded nuts using impulses rather than continuous force.

Example 2: A warning light flashes so that it is even more noticeable than when consistently lit.

These are just examples of the 40 Inventive Principles. You need to become familiar with all 40 and how to use them; however, due to the limited space in this book devoted to TRIZ, the reader will need to further investigate the TRIZ methodology by taking a class or reading the suggested documents. The following are additional suggestions that the reader should study related to the TRIZ methodology in order to be effective in using it:

- HU diagrams
- I-TRIZ
- Separation principles
- Substance–field analysis
- System of standard solutions
- Technological effects and phenomena

SOFTWARE

There are a number of software packages available. Some are very complex, while others are very simple and easy to learn but do not provide all the options that some of the more expensive software packages include. I have been using Ideation International software packages and have been satisfied with their performance. It does not mean that it is the best one for you. I do recommend that you select one of the software packages as it greatly reduces cycle time to come up with an excellent/creative solution.

Ideation International is the leading provider of innovation software based on TRIZ and the TRIZSoft® theoretical foundation.

The following TRIZSoft products are currently available:

- Innovation WorkBench®—A comprehensive professional tool for inventive problem solving.
- Ideation Brainstorming—A simplified tool for solving problems of light to medium complexity in individual or team work format.

- Knowledge Wizard—A professional tool for inventive problem solving in nontechnical areas (business, management, marketing, logistics, etc.).
- Failure Analysis—A professional tool for revealing root causes of undesired effects (accidents, failures, production defects, etc.) and their elimination (prevention).
- Failure Prediction—A professional tool for predicting possible undesired effects and events (accidents, failures, production defects, etc.) and their prevention.
- Intellectual Property Management—A professional tool for evaluating and enhancing patents, disclosures, and applications, *inventing around* patents, and protecting patents from inventive competitors.
- Ideation Brainstorming for solving nontechnical problems—A tool for solving problems in business, management, marketing, logistics, etc., in an individual or team setting.
- IDM-TRIZ powered by STEPS!—A powerful method developed to guide the engineer in his efforts to innovate. The method has a series of steps guiding the engineer in a creative design process (http://www.time-to-innovate.com/en).

REFERENCES

Altshuller, G. *40 Principles: TRIZ Keys to Technical Innovation.* Translated by L. Shulyak and S. Rodman. Worchester, Massachusetts: Technical Innovation Center, 1997. 141 pages. ISBN 0964074036.

Bukhman, I. *TRIZ Technology for Innovation.* Isak Bukhman, Taipei, Taiwan: Cubic Creativity Company. 2012. 368 pages. ISBN 978-986-85635-2-0.

SUGGESTED ADDITIONAL READING

Altshuller, G. *And Suddenly the Inventor Appeared: TRIZ, the Theory of Inventive Problem Solving.* Worchester, MA: Technical Innovation Center, 1996. ISBN 0-9640740-2-8.

Altshuller, G. *Creativity as an Exact Science: The Theory of the Solution of Inventive Problems.* Translated by A. Williams. NY: Gordon and Breach Science Publishers, 1984. ISBN 0-677-21230-5.

Altshuller, G. *The Innovation Algorithm: TRIZ, Systematic Innovation, and Technical Creativity.* Worchester, MA: Technical Innovation Center, 1999. 312 pages. ISBN 0964074044.

Clarke Sr., D.W. *TRIZ: Through the Eyes of an American TRIZ Specialist*. Detroit, Farmington, MI: Ideation International Inc., 1997. ISBN 1928747035.

Fey, V.R. and Rivin, E.I. *The Science of Innovation: A Managerial Overview of the TRIZ Methodology*. West Bloomfield, MI: The TRIZ Group, 1997. 82 pages. ISBN 0-9658359-0-1.

Kaplan, S. *An Introduction to TRIZ*. Detroit, Farmington, MI: Ideation International Inc., 1996. ISBN 1928747000.

Kosse, V. *Solving Problems with TRIZ: An Exercise Handbook*. Detroit, Farmington, MI: Ideation International Inc., 1999. 123 pages. ISBN 1928747019.

Savransky, S.D. *Engineering of Creativity: Introduction to TRIZ Methodology of Inventive Problem Solving*. NY: CRC Press. 2000. 394 pages. ISBN 0849322553.

Terninko, J., Zusman, A., and Zlotin, B. *Step-by-Step TRIZ: Creating Innovative Solution Concepts*. NY: CRC Press. 1997.

Terninko, J., Zusman, A., and Zlotin, B. *Systematic Innovation: An Introduction to TRIZ (Theory of Inventive Problem Solving)*. 1998. 150 pages. ISBN 1574441116.

Zlotin, B. and Zusman, A. *Directed Evolution: Philosophy, Theory and Practice*. Detroit: Ideation International Inc., 2001. 103 pages.

Zlotin, B., Zusman, A., Altshuller, G., Philatov, V. *Tools of Classical TRIZ*. Detroit: Ideation International Inc., 1999. 266 pages, ISBN 1928747027.

CONTRADICTION TABLES

The combinations of Figures 29.5 through 29.10 make up the total TRIZ table.

Contradiction Table

Feature to improve \ Undesired result (conflict)	1 Weight of moving object	2 Weight of nonmoving object	3 Length of moving object	4 Length of nonmoving object	5 Area of moving object	6 Area of nonmoving object	7 Volume of moving object	8 Volume of nonmoving object	9 Speed	10 Force	11 Tension, pressure	12 Shape	13 Stability of object
1 Weight of moving object			15, 8, 29, 34		29, 17, 38, 34		29, 2, 40, 28		2, 8, 15, 38	8, 10, 18, 37	10, 36, 37, 40	10, 14, 35, 40	1, 35, 19, 39
2 Weight of nonmoving object				10, 1, 29, 35		35, 30, 13, 2		5, 35, 14, 2		8, 10, 19, 35	13, 29, 10, 18	13, 10, 29, 14	26, 39, 1, 40
3 Length of moving object	8, 15, 29, 34				15, 17, 4		7, 17, 4, 35		13, 4, 8	17, 10, 4	1, 8, 35	1, 8, 10, 29	1, 8, 15, 34
4 Length of nonmoving object		35, 28, 40, 29					17, 7, 10, 40		35, 8, 2, 14	28, 10	1, 14, 35	13, 14, 15, 7	39, 37, 35
5 Area of moving object	2, 17, 29, 4		14, 15, 18, 4				7, 14, 17, 4		29, 30, 4, 34	19, 30, 35, 2	10, 15, 36, 28	5, 34, 29, 4	11, 2, 13, 39
6 Area of nonmoving object		30, 2, 14, 18		26, 7, 9, 39						1, 8, 35, 36	10, 15, 36, 37		2, 38
7 Volume of moving object	2, 26, 29, 40		1, 7, 4, 35		1, 7, 4, 17				29, 4, 38, 34	15, 35, 36, 37	6, 35, 36, 37	1, 15, 29, 4	28, 10, 1, 39
8 Volume of nonmoving object		35, 10, 19, 14	19, 14	35, 8, 2, 14						2, 18, 37	24, 35	7, 2, 35	34, 28, 35, 40
9 Speed	2, 28, 13, 38		13, 14, 8		29, 30, 34		7, 29, 34			13, 28, 15, 19	6, 18, 38, 40	35, 15, 18, 34	28, 33, 1, 18
10 Force	8, 1, 37, 18	18, 13, 1, 28	17, 19, 6, 36	28, 10	19, 10, 15	1, 18, 36, 37	15, 9, 12, 37	2, 36, 18, 37	13, 28, 15, 12		18, 21, 11	10, 35, 40, 34	35, 10, 21
11 Tension, pressure	10, 36, 37, 40	13, 29, 10, 18	35, 10, 36	35, 1, 14, 16	10, 15, 36, 25	10, 15, 35, 37	6, 35, 10	35, 24	6, 35, 36	36, 35, 21		35, 4, 15, 10	35, 33, 2, 40
12 Shape	8, 10, 29, 40	15, 10, 26, 3	29, 34, 5, 4	13, 14, 10, 7	5, 34, 4, 10		14, 4, 15, 22	7, 2, 35	35, 15, 34, 18	35, 10, 37, 40	34, 15, 10, 14		33, 1, 18, 4
13 Stability of object	21, 35, 2, 39	26, 39, 1, 40	13, 15, 1, 28	37	2, 11, 13	39	28, 10, 19, 39	34, 28, 35, 40	33, 15, 28, 18	10, 35, 21, 16	2, 35, 40	22, 1, 18, 4	
14 Strength	1, 8, 40, 15	40, 26, 27, 1	1, 15, 8, 35	15, 14, 28, 26	3, 34, 40, 29	9, 40, 28	10, 15, 14, 7	9, 14, 17, 15	8, 13, 26, 14	10, 18, 3, 14	10, 3, 18, 40	10, 30, 35, 40	13, 17, 35
15 Durability of moving object	19, 5, 34, 31		2, 19, 9		3, 17, 19		10, 2, 19, 30		3, 35, 5	19, 2, 16	19, 3, 27	14, 26, 28, 25	13, 3, 35
16 Durability of nonmoving object		6, 27, 19, 16		1, 10, 35				35, 34, 38					39, 3, 35, 23
17 Temperature	36, 22, 6, 38	22, 35, 32	15, 19, 9	15, 19, 9	3, 35, 39, 18	35, 38	34, 39, 40, 18	35, 6, 4	2, 28, 36, 30	35, 10, 3, 21	35, 39, 19, 2	14, 22, 19, 32	1, 35, 32
18 Brightness	19, 1, 32	2, 35, 32			19, 32, 26								
19 Energy spent by moving object	12, 18, 28, 31		12, 28		15, 19, 25		35, 13, 18		8, 15, 35	16, 26, 21, 2	23, 14, 25	12, 2, 29	19, 13, 17, 24
20 Energy spent by nonmoving object		19, 9, 6, 27								36, 37			27, 4, 29, 19

FIGURE 29.5

Part 1 of the contradiction table.

Contradiction Table

Feature to improve		14 Strength	15 Durability of moving object	16 Durability of nonmoving object	17 Temperature	18 Brightness	19 Energy spent by moving object	20 Energy spent by nonmoving object	21 Power	22 Waste of energy	23 Waste of substance	24 Loss of information	25 Waste of time	26 Amount of substance
1	Weight of moving object	28, 27, 18, 40	5, 34, 31, 35		6, 20, 4, 38	19, 1, 32	35, 12, 34, 31		12, 36, 18, 31	6, 2, 34, 19	5, 35, 3, 31	10, 24, 35	10, 35, 20, 28	3, 26, 18, 31
2	Weight of nonmoving object	28, 2, 10, 27		2, 27, 19, 6	28, 19, 32, 22	19, 32, 35		18, 19, 28, 1	15, 19, 18, 22	18, 19, 28, 15	5, 8, 13, 30	10, 15, 35	10, 20, 35, 26	19, 6, 18, 26
3	Length of moving object	8, 35, 29, 34	19		10, 15, 19	32	8, 35, 24		1, 35	7, 2, 35, 39	4, 29, 23, 10	1, 24	15, 2, 29	29, 35
4	Length of nonmoving object	15, 14, 28, 26		1, 40, 35	3, 35, 38, 18	3, 25			12, 8	6, 28	10, 28, 24, 35	24, 26	30, 29, 14	
5	Area of moving object	3, 15, 40, 14	6, 3		2, 15, 16	15, 32, 19, 13	19, 32		19, 10, 32, 18	15, 17, 30, 26	10, 35, 2, 39	30, 26	26, 4	29, 30, 6, 13
6	Area of nonmoving object	40		2, 10, 19, 30	35, 39, 38				17, 32	17, 7, 30	10, 14, 18, 39	30, 16	10, 35, 4, 18	2, 18, 40, 4
7	Volume of moving object	9, 14, 15, 7	6, 35, 4		34, 39, 10, 18	2, 13, 10	35		35, 6, 13, 18	7, 15, 13, 16	36, 39, 34, 10	2, 22	2, 6, 34, 10	29, 30, 7
8	Volume of nonmoving object	9, 14, 17, 15		35, 34, 38	35, 6, 4				30, 6		10, 39, 35, 34		35, 16, 32, 18	35, 3
9	Speed	8, 3, 26, 14	3, 19, 35, 5		28, 30, 36, 2	10, 13, 19	8, 15, 35, 38		19, 35, 38, 2	14, 20, 19, 35	10, 13, 28, 38	13, 26		18, 19, 29, 38
10	Force	35, 10, 14, 27	19, 2		35, 10, 24		19, 17, 10	1, 16, 36, 37	19, 35, 18, 37	14, 15	8, 35, 40, 5		10, 37, 36	14, 29, 18, 36
11	Tension, pressure	9, 18, 3, 40	19, 3, 27		35, 39, 19, 2		14, 24, 10, 37		10, 35, 14	2, 36, 25	10, 36, 3, 37		37, 36, 4	10, 14, 36
12	Shape	30, 14, 10, 40	14, 26, 9, 25		22, 14, 19, 32	13, 15, 32	2, 6, 34, 14		4, 6, 2	14	35, 29, 3, 5		14, 10, 34, 17	36, 22
13	Stability of object	17, 9, 15	13, 27, 10, 35	39, 3, 35, 23	35, 1, 32	32, 3, 27, 15	13, 19	27, 4, 29, 18	32, 35, 27, 31	14, 2, 39, 6	2, 14, 30, 40		35, 27	15, 32, 35
14	Strength		27, 3, 26		30, 10, 40	35, 19, 10	19, 35, 10	35	10, 26, 35, 28	35	35, 28, 31, 40		29, 3, 28, 10	29, 10, 27
15	Durability of moving object	27, 3, 10			19, 35, 39	2, 19, 4, 35	28, 6, 35, 18		19, 10, 35, 38		28, 27, 3, 18	10	20, 10, 28, 18	3, 35, 10, 40
16	Durability of nonmoving object				19, 18, 36, 40				16		27, 16, 18, 38	10	28, 20, 10, 16	3, 35, 31
17	Temperature	10, 30, 22, 40	19, 13, 39	19, 18, 36, 40		32, 30, 21, 16	19, 15, 3, 17		2, 14, 17, 25	21, 17, 35, 38	21, 36, 29, 31		35, 28, 21, 18	3, 17, 30, 39
18	Brightness	35, 19	2, 19, 6		32, 35, 19		32, 1, 19	32, 35, 1, 15	32	19, 16, 1, 6	13, 1	1, 6	19, 1, 26, 17	1, 19
19	Energy spent by moving object	5, 19, 9, 35	28, 35, 6, 18		19, 24, 3, 14	2, 15, 19			6, 19, 37, 18	12, 22, 15, 24	35, 24, 18, 5		35, 38, 19, 18	34, 23, 16, 18
20	Energy spent by nonmoving object	35				19, 2, 35, 32					28, 27, 18, 31			3, 35, 31

FIGURE 29.6

Part 2 of the contradiction table.

Contradiction Table

Feature to improve	Undesired result (conflict)	27 Reliability	28 Accuracy of measurement	29 Accuracy of manufacturing	30 Harmful factors acting on object	31 Harmful side effects	32 Manufacturability	33 Convenience of use	34 Repairability	35 Adaptability	36 Complexity of device	37 Complexity of control	38 Level of automation	39 Productivity
1	Weight of moving object	3, 11, 1, 27	28, 27, 35, 26	28, 35 26, 18	22, 21, 18, 27	22, 35, 31,39	27, 28, 1, 36	35, 3, 2, 24	2, 27, 28, 11	29, 5, 15, 8	26, 30, 36, 34	28, 29, 26, 32	26, 35, 18, 19	35, 3, 24, 37
2	Weight of nonmoving object	10, 28, 8, 3	18, 26, 28	10, 1, 35, 17	2, 19, 22, 37	35, 22, 1,39	28, 1, 9	6, 13, 1, 32	2, 27, 28, 11	19, 15, 29	1, 10, 26, 39	25, 28, 17, 15	2, 26, 35	1, 28, 15, 35
3	Length of moving object	10, 14, 29, 40	28, 32, 4	10, 28, 29, 37	1, 15, 17, 24	17, 15	1, 29, 17	15, 29, 35, 4	1, 28, 10	14, 15, 1, 16	1, 19, 26, 24	35, 1, 26, 24	17, 24, 26, 16	14, 4, 28, 29
4	Length of nonmoving object	15, 29, 28	32, 28, 3	2, 32, 10	1, 18		15, 17, 27	2, 25	3	1, 35	1, 26	26		30, 14, 7, 26
5	Area of moving object	29, 9	26, 28, 32, 3	2, 32	22, 33, 28, 1	17, 2, 18, 39	13, 1, 26, 24	15, 17, 13, 16	15, 13, 10, 1	15, 30	14, 1, 13	2, 36, 26, 18	14, 30, 28, 23	10, 26, 34, 2
6	Area of nonmoving object	32, 35, 40, 4	26, 28, 32, 3	2, 29, 18, 36	27, 2, 39, 35	22, 1, 40	40, 16	16, 4	16	15, 16	1, 18, 36	2, 35, 30, 18	23	10, 15, 17, 7
7	Volume of moving object	14, 1, 40, 11	25, 26, 28	25, 28, 2, 16	22, 21, 27, 35	17, 2, 40, 1	29, 1, 40	15, 13, 30, 12	10	15, 29	26, 1	29, 26, 4	35, 34, 16, 24	10, 6, 2, 34
8	Volume of nonmoving object	2, 35, 16		35, 10, 25	34, 39, 19, 27	30, 18, 35, 4	35		1		1, 31	2, 17, 26		35, 37, 10, 2
9	Speed	11, 35, 27, 28	28, 32, 1, 24	10, 28, 32, 25	1, 28, 35, 23	2, 24, 35, 21	35, 13, 8, 1	32, 28, 13, 12	34, 2, 28, 27	15, 10, 26	10, 28, 4, 34	3, 34, 27, 16	10, 18	
10	Force	3, 35, 13, 21	35, 10, 23, 24	28, 29, 37, 36	1, 35, 40, 18	13, 3, 36, 24	15, 37, 18, 1	1, 28, 3, 25	15, 1, 11	15, 17, 18, 20	26, 35, 10, 18	36, 37, 10, 19	2, 35	3, 28, 35, 37
11	Tension, pressure	10, 13, 19, 35	6, 28, 25	3, 35	22, 2, 37	2, 33, 27, 18	1, 35, 16	11	2	35	19, 1, 35	2, 36, 37	35, 24	10, 14, 35, 37
12	Shape	10, 40, 16	28, 32, 1	32, 30, 40	22, 1, 2, 35	35, 1	1, 32, 17, 28	32, 15, 26	2, 13, 1	1, 15, 29	16, 29, 1, 28	15, 13, 39	15, 1, 32	17, 26, 34, 10
13	Stability of object		13	18	35, 24, 30, 18	35, 40, 27, 39	35, 19	32, 35, 30	2, 35, 10, 16	35, 30, 34, 2	2, 35, 22, 26	35, 22, 21, 35	1, 8, 35	23, 35, 40, 3
14	Strength	11, 3	3, 27, 16	3, 27	18, 35, 37, 1	15, 35, 22, 2	11, 3, 10, 32	32, 40, 28, 2	27, 11, 3	15, 3, 32	2, 13, 28	27, 3, 15, 40	15	29, 35, 10, 14
15	Durability of moving object	11, 2, 13	3	3, 27, 16, 40	22, 15, 33, 28	21, 39, 16, 22	27, 1, 4	12, 27	29, 10, 27	1, 35, 13	10, 4, 29, 15	19, 29, 39, 35	6, 10	35, 17, 14, 19
16	Durability of nonmoving object	34, 27, 6, 40	10, 26, 24		17, 1, 40, 33	22	35, 10	1	1	2		25, 34, 6, 35	1	10, 20, 16, 38
17	Temperature	19, 35, 3, 10	32, 19, 24	24	22, 33, 35, 2	22, 35, 2, 24	26, 27	26, 27	4, 10, 16	2, 18, 27	2, 17, 16	3, 27, 35, 31	26, 2, 19, 16	15, 28, 35
18	Brightness		11, 15, 32	3, 32	15, 19	35, 19, 32, 39	19, 35, 28, 26	28, 26, 19	15, 17, 13, 16	15, 1, 1, 19	6, 32, 13	32, 15	2, 26, 10	2, 25, 16
19	Energy spent by moving object	19, 21, 11, 27	3, 1, 32		1, 35, 6, 27	2, 35, 6	28, 26, 30	19, 35	1, 15, 17, 28	15, 17, 13, 16	2, 29, 27, 28	35, 38	32, 2	12, 28, 35
20	Energy spent by nonmoving object	10, 36, 23			10, 2, 22, 37	19, 22, 18	1, 4					19, 35, 16, 25		1, 6

FIGURE 29.7

Part 3 of the contradiction table.

Contradiction Table

Feature to improve		1 Weight of moving object	2 Weight of nonmoving object	3 Length of moving object	4 Length of nonmoving object	5 Area of moving object	6 Area of nonmoving object	7 Volume of moving object	8 Volume of nonmoving object	9 Speed	10 Force	11 Tension, pressure	12 Shape	13 Stability of object
21	Power	8, 36, 38, 31	19, 26, 17, 27	1, 10, 35, 37		19, 38	17, 32, 13, 38	35, 6, 38	30, 6, 25	15, 35, 2	26, 2, 36, 35	22, 10, 35	29, 14, 2, 40	35, 32, 15, 31
22	Waste of energy	15, 6, 19, 28	19, 6, 18, 9	7, 2, 6, 13	6, 38, 7	15, 26, 17, 30	17, 7, 30, 18	7, 18, 23	7	16, 35, 38	36, 38			14, 2, 39, 6
23	Waste of substance	35, 6, 23, 40	35, 6, 22, 32	14, 29, 10, 398	10, 28, 24	35, 2, 10, 31	10, 18, 39, 31	1, 29, 30, 36	3, 39, 18, 31	10, 13, 28, 38	14, 15, 18, 40	3, 36, 37, 10	29, 35, 3, 5	2, 14, 30, 40
24	Loss of information	10, 24, 35	10, 35, 5	1, 26	26	30, 26	30, 16		2, 22	26, 32				
25	Waste of time	10, 20, 37, 35	10, 20, 26, 5	15, 2, 29	30, 24, 14, 5	26, 4, 5, 16	10, 35, 17, 4	2, 5, 34, 10	35, 16, 32, 18		10, 37, 36, 5	37, 36, 4	4, 10, 34, 17	35, 3, 22, 5
26	Amount of substance	35, 6, 18, 31	27, 26, 18, 35	29, 14, 35, 18		15, 14, 29	2, 18, 40, 4	15, 20, 29		35, 29, 34, 28	35, 14, 3	10, 36, 14, 3	35, 14	15, 2, 17, 40
27	Reliability	3, 8, 10, 40	3, 10, 8, 28	15, 9, 14, 4	15, 29, 28, 11	17, 10, 14, 16	32, 35, 40, 4	3, 10, 14, 24	2, 35, 24	21, 35, 11, 28	8, 28, 10, 3	10, 24, 35, 19	35, 1, 16, 11	
28	Accuracy of measurement	32, 35, 26, 28	28, 35, 25, 26	28, 26, 5, 16	32, 28, 3, 16	26, 28, 32, 3	26, 28, 32, 3	32, 13, 6		28, 13, 32, 24	32, 2	6, 28, 32	6, 28, 32	32, 35, 13
29	Accuracy of manufacturing	28, 32, 13, 18	28, 35, 27, 9	10, 28, 29, 37	2, 32, 10	28, 33, 29, 32	2, 29, 18, 36	32, 28, 2	25, 10, 35	10, 28, 32	28, 19, 34, 36	3, 35	32, 30, 40	30, 18
30	Harmful factors acting on object	22, 21, 27, 39	2, 22, 13, 24	17, 1, 39, 4	1, 18	22, 1, 33, 28	27, 2, 39, 35	22, 23, 37, 35	34, 39, 19, 27	21, 22, 35, 28	13, 35, 39, 18	22, 2, 37	22, 1, 3, 35	35, 24, 30, 18
31	Harmful side effects	19, 22, 15, 39	35, 22, 1, 39	17, 15, 16, 22		17, 2, 18, 39	22, 1, 40	17, 2, 40	30, 18, 35, 4	35, 28, 3, 23	35, 28, 1, 40	2, 33, 27, 18	35, 1	35, 40, 27, 39
32	Manufacturability	28, 29, 15, 16	1, 27, 36, 13	1, 29, 13, 17	15, 17, 27	13, 1, 26, 12	16, 40	13, 29, 1, 40	35	35, 13, 8, 1	35, 12	35, 19, 1, 37	1, 28, 13, 27	11, 13, 1
33	Convenience of use	25, 2, 13, 15	6, 13, 1, 25	1, 17, 13, 12		1, 17, 13, 16	18, 16, 15, 39	1, 16, 35, 15	4, 18, 39, 31	18, 13, 34	28, 13, 35	2, 32, 12	15, 34, 29, 28	32, 35, 30
34	Repairability	2, 27, 35, 11	2, 27, 35, 11	1, 28, 10, 25	3, 18, 31	15, 13, 32	16, 25	25, 2, 35, 11	1	34, 9	1, 11, 10	13	1, 13, 2, 4,	2, 35
35	Adaptability	1, 6, 15, 8	19, 15, 29, 16	35, 1, 29, 2	1, 35, 16	35, 30, 29, 7	15, 16	15, 35, 29		35, 10, 14	15, 17, 20	35, 16	15, 37, 1, 8	35, 30, 14
36	Complexity of device	26, 30, 34, 36	2, 36, 35, 39	1, 19, 26, 24	26	14, 1, 13, 16	6, 36	34, 25, 6	1, 16	34, 10, 28	26, 16	19, 1, 35	29, 13, 28, 15	2, 22, 17, 19
37	Complexity of control	27, 26, 28, 13	6, 13, 28, 1	16, 17, 26, 24	26	2, 13, 15, 17	2, 39, 30, 16	29, 1, 4, 16	2, 18, 26, 31	3, 4, 16, 35	36, 28, 40, 19	35, 36, 37, 32	27, 13, 1, 39	11, 22, 39, 30
38	Level of automation	28, 26, 18, 35	28, 26, 35, 10	14, 13, 17, 28	23	17, 14, 13		35, 13, 16		28, 10	2, 35	13, 35	15, 32, 1, 13	18, 1
39	Productivity	35, 26, 24, 37	28, 27, 15, 3	18, 4, 28, 38	30, 7, 14, 26	10, 26, 34, 31	10, 35, 17, 7	2, 6, 34, 10	35, 37, 10, 2		28, 15, 10, 36	10, 37, 14	14, 10, 34, 40	35, 3, 22, 39

FIGURE 29.8

Part 4 of the contradiction table.

Contradiction Table

Feature to improve		14 Strength	15 Durability of moving object	16 Durability of nonmoving object	17 Temperature	18 Brightness	19 Energy spent by moving object	20 Energy spent by nonmoving object	21 Power	22 Waste of energy	23 Waste of substance	24 Loss of information	25 Waste of time	26 Amount of substance
21	Power	26, 10, 28	19, 35, 10, 38	16	2, 14, 17, 25	16, 6, 19	16, 6, 19, 37			10, 35, 38	28, 27, 18, 38	10, 19	35, 20, 10, 6	4, 34, 19
22	Waste of energy	26			19, 38, 7	1, 13, 32, 15			3, 38		35, 27, 2, 37	19, 10	10, 18, 32, 7	7, 18, 25
23	Waste of substance	35, 28, 31, 40	28, 27, 3, 18	27, 16, 18, 38	21, 36, 39, 31	1, 6, 13	35, 18, 24, 5	28, 27, 12, 31	28, 27, 18, 38	35, 27, 2, 31			15, 18, 35, 10	6, 3, 10, 24
24	Loss of information		10	10		19			10, 19	10, 19			24, 26, 28, 32	24, 28, 35
25	Waste of time	29, 3, 28, 18	20, 10, 28, 18	28, 20, 10, 16	35, 29, 21, 18	1, 19, 26, 17	35, 38, 19, 18	1	35, 20, 10, 6	10, 5, 18, 32	35, 18, 10, 39	24, 26, 28, 32		35, 38, 18, 16
26	Amount of substance	14, 35, 34, 10	3, 35, 10, 40	3, 35, 31	3, 17, 39		34, 29, 16, 18	3, 35, 31	35	7, 18, 25	6, 3, 10, 24	24, 28, 35	35, 38, 18, 16	
27	Reliability	11, 28	2, 35, 3, 25	34, 27, 6, 40	3, 35, 10	11, 32, 13	21, 11, 27, 19	36, 23	21, 11, 26, 31	10, 11, 35	10, 35, 29, 39	10, 28	10, 30, 4	21, 28, 40, 3
28	Accuracy of measurement	28, 6, 32	28, 6, 32	10, 26, 24	6, 19, 28, 24	6, 1, 32	3, 6, 32		3, 6, 32	26, 32, 27	10, 16, 31, 28		24, 34, 28, 32	2, 6, 32
29	Accuracy of manufacturing	3, 27	3, 27, 40		19, 26	3, 32	32, 2		32, 2	13, 32, 2	35, 31, 10, 24		32, 26, 28, 18	32, 30
30	Harmful factors acting on object	18, 35, 37, 1	22, 15, 33, 28	17, 1, 40, 33	22, 33, 35, 2	1, 19, 32, 13	1, 24, 6, 27	10, 2, 22, 37	19, 22, 31, 2	21, 22, 35, 2	33, 22, 19, 40	22, 10, 2	35, 18, 34	35, 33, 29, 31
31	Harmful side effects	15, 35, 22, 2	15, 22, 33, 31	21, 39, 16, 22	22, 35, 2, 24	19, 24, 39, 32	2, 35, 6	19, 22, 18	2, 35, 18	21, 35, 2, 22	10, 1, 34	10, 21, 29	1, 22	3, 24, 39, 1
32	Manufacturability	1, 3, 10, 32	27, 1, 4	35, 16	27, 26, 18	28, 24, 27, 1	28, 26, 27, 1	1, 4	27, 1, 12, 24	19, 35	15, 34, 33	32, 24, 18, 16	35, 28, 34, 4	35, 23, 1, 24
33	Convenience of use	32, 40, 3, 28	29, 3, 8, 25	1, 16, 25	26, 27, 13	13, 17, 1, 24	1, 13, 24		35, 34, 2, 10	2, 19, 13	28, 32, 2, 24	4, 10, 27, 22	4, 28, 10, 34	12, 35
34	Repairability	11, 1, 2, 9	11, 29, 28, 27	1	4, 10	15, 1, 13	15, 1, 28, 16		15, 10, 32, 2	15, 1, 32, 19	2, 35, 34, 27		32, 1, 10, 25	2, 28, 10, 25
35	Adaptability	35, 3, 32, 6	13, 1, 35	2, 16	27, 2, 3, 35	6, 22, 26, 1	19, 35, 29, 13		19, 1, 29	18, 15, 1	15, 10, 2, 13		35, 28	3, 35, 15
36	Complexity of device	2, 13, 28	10, 4, 28, 15		2, 17, 13	24, 17, 13	27, 2, 29, 28		20, 19, 30, 34	10, 35, 13, 2	35, 10, 28, 29		6, 29	13, 3, 27, 10
37	Complexity of control	27, 3, 15, 28	19, 29, 39, 25	25, 24, 6, 35	3, 27, 35, 16	2, 24, 26	35, 38	19, 35, 16	19, 1, 16, 10	35, 3, 15, 19	1, 13, 10, 24	35, 33, 27, 22	18, 28, 32, 9	3, 27, 29, 18
38	Level of automation	25, 13	6, 9		26, 2, 19	8, 32, 19	2, 32, 13		28, 2, 27	23, 28	35, 10, 18, 5	35, 33	24, 28, 35, 30	35, 13
39	Productivity	29, 28, 10, 18	35, 10, 2, 18	20, 10, 16, 38	35, 21, 28, 10	26, 17, 19, 1	35, 10, 38, 19	1	35, 20, 10	28, 10, 29, 35	28, 10, 35, 23	13, 15, 23		35, 38

FIGURE 29.9

Part 5 of the contradiction table.

Contradiction Table

		27	28	29	30	31	32	33	34	35	36	37	38	39
Feature to improve	Undesired result (conflict)	Reliability	Accuracy of measurement	Accuracy of manufacturing	Harmful factors acting on object	Harmful side effects	Manufacturability	Convenience of use	Repairability	Adaptability	Complexity of device	Complexity of control	Level of automation	Productivity
21	Power	19, 24, 26, 31	32, 15, 2	32, 2	19, 22, 31, 2	2, 35, 18	26, 10, 34	26, 35, 10	35, 2, 10, 34	19, 17, 34	20, 19, 30, 34	19, 35, 16	28, 2, 17	28, 35, 34
22	Waste of energy	11, 10, 35	32		21, 22, 35, 2	21, 35, 2, 22		35, 22, 1	2, 19		7, 23	35, 3, 15, 23	2	28, 10, 29, 35
23	Waste of substance	10, 29, 39, 35	16, 34, 31, 28	35, 10, 24, 31	33, 22, 30, 40	10, 1, 34, 29	15, 34, 33	32, 28, 2, 24	2, 35, 34, 27	15, 10, 2	35, 10, 28, 24	35, 18, 10, 13	35, 10, 18	28, 35, 10, 23
24	Loss of information	10, 28, 23			22, 10, 1	10, 21, 22	32	27, 22				35, 33	35	13, 23, 15
25	Waste of time	10, 30, 4	24, 34, 28, 32	24, 26, 28, 18	35, 18, 34	35, 22, 18, 39	35, 28, 34, 4	4, 28, 10, 34	32, 1, 10	35, 28	6, 29	18, 28, 32, 10	24, 28, 35, 30	
26	Amount of substance	18, 3, 28, 40	13, 2, 28	33, 30	35, 33, 29, 31	3, 35, 40, 39	29, 1, 35, 27	35, 29, 25, 10	2, 32, 10, 25	15, 3, 29	3, 13, 27, 10	3, 27, 29, 18	8, 35	13, 29, 3, 27
27	Reliability		32, 3, 11, 23	11, 32, 1	27, 35, 2, 40	35, 2, 40, 26		27, 17, 40	1, 11	13, 35, 8, 24	13, 35, 1	27, 40, 28	11, 13, 27	1, 35, 29, 38
28	Accuracy of measurement	5, 11, 1, 23			28, 24, 22, 26	3, 33, 39, 10	6, 35, 25, 18	1, 13, 17, 34	1, 32, 13, 11	13, 35, 2	27, 35, 10, 34	26, 24, 32, 28	28, 2, 10, 34	10, 34, 28, 32
29	Accuracy of manufacturing	11, 32, 1			26, 28, 10, 36	4, 17, 34, 26		1, 32, 35, 23	25, 10		26, 2, 18		26, 28, 18, 23	10, 18, 32, 39
30	Harmful factors acting on object	27, 24, 2, 40	28, 33, 23, 26	26, 28, 10, 18			24, 35, 2	2, 25, 28, 39	35, 10, 2	35, 11, 22, 31	22, 19, 29, 40	22, 19, 29, 40	33, 3, 34	22, 35, 13, 24
31	Harmful side effects	24, 2, 40, 39	3, 33, 26	4, 17, 34, 26							19, 1, 31	2, 21, 27, 1	2	22, 35, 18, 39
32	Manufacturability		1, 35, 12, 18		24, 2			2, 5, 13, 16	35, 1, 11, 9	2, 13, 15	27, 26, 1	6, 28, 11, 1	8, 28, 1	35, 1, 10, 28
33	Convenience of use	17, 27, 8, 40	25, 13, 2, 34	1, 32, 35, 23	2, 25, 28, 39		2, 5, 12		12, 26, 1, 32	15, 34, 1, 16	32, 26, 12, 17		1, 34, 12, 3	15, 1, 28
34	Repairability	11, 10, 1, 16	10, 2, 13	25, 10	35, 10, 2, 16		1, 35, 11, 10	1, 12, 26, 15		7, 1, 4, 16	35, 1, 13, 11		34, 35, 7, 13	1, 32, 10
35	Adaptability	35, 13, 8, 24	35, 5, 1, 10		35, 11, 32, 31		1, 13, 31	15, 34, 1, 16	1, 16, 7, 4		15, 29, 37, 28	1	27, 34, 35	35, 28, 6, 37
36	Complexity of device	13, 35, 1	2, 26, 10, 34	26, 24, 32	22, 19, 29, 40	19, 1	27, 26, 1, 13	27, 9, 26, 24	1, 13	29, 15, 28, 37		15, 10, 37, 28	15, 1, 24	12, 17, 28
37	Complexity of control	27, 40, 28, 8	26, 24, 32, 28		22, 19, 29, 28	2, 21	5, 28, 11, 29	2, 5	12, 26	1, 15	15, 10, 37, 28		34, 21	35, 18
38	Level of automation	11, 27, 32	28, 26, 10, 34	28, 26, 18, 23	2, 33	2	1, 26, 13	1, 12, 34, 3	1, 35, 13	27, 4, 1, 35	15, 24, 10	34, 27, 25		5, 12, 35, 26
39	Productivity	1, 35, 10, 38	1, 10, 34, 28	18, 10, 32, 1	22, 35, 13, 24	32, 22, 18, 39	35, 28, 2, 24	1, 28, 7, 19	1, 32, 10, 25	1, 35, 28, 37	12, 17, 28, 24	35, 18, 27, 2	5, 12, 35, 26	

FIGURE 29.10

Part 6 of the contradiction table.

Appendix: Innovation Definitions

The following terms and definitions are all direct quotes from the International Association of Innovation Professionals (IAOIP) "Study Guide to the Basic Certification Exam." The study guide is used to prepare individuals to take the basic examination to be certified as a professional innovator by the IAOIP. These were taken from what the IAOIP is using as the body of knowledge for the innovation professional.

Notes:

1. The terms and definitions that are italicized were not included in the "Study Guide to the Basic Certification Exam" but are included in the methodologies and tools documented technology.
2. In some cases, there is more than one definition for the same tool/methodology. In most of these cases, all the definitions are acceptable and often the additional definitions just help to clarify what the methodology/tool is. In some cases, the preferred definition is identified.

TERMS AND DEFINITIONS

5 Whys: *A simple but effective method of asking five times why a problem occurred. After each answer, ask why again using the previous response. It is surprising how this may lead to a root cause of the problem, but it does not solve the problem.*

5 Whys: A technique to get to the root cause of the problem. It is the practice of asking five times or more why the failure has occurred in order to get to the root cause. Each time an answer is given, you ask why that particular condition occurred. As outlined in the 5 Whys Overview, it is recommended that the 5 Whys be used with risk assessment in order to strengthen the use of the tool for innovation and creativity-enhancing purposes.

76 Standard Solutions: *A collection of problem-solving concepts intended to help innovators develop solutions. A list was developed from referenced works and published in a comparison with the 40 principles to show that those who are familiar with the 40 principles will be able to expand their problem-solving capability. They are grouped into five categories as follows:*
1. *Improving the system with no or little change: 13 standard solutions*
2. *Improving the system by changing the system: 23 standard solutions*
3. *System transitions: 6 standard solutions*
4. *Detection and measurement: 17 standard solutions*
5. *Strategies for simplification and improvement: 17 standard solutions*

7–14–28 Processes: This is a task-analysis assessment that involves breaking a process down into 7 tasks, then breaking it further into 14 tasks, and then another level further into 28 tasks.

40 Inventive Principles: The 40 Inventive Principles that form a core part of the TRIZ (theory of inventive problem solving) methodology invented by G.S. Altshuller. These are 40 tools used to overcome technical contradictions. Each is a generic suggestion for performing an action to, and within, a technical system. For example principle #1 (segmentation) suggests finding a way to separate two elements of a technical system into many small interconnected elements.

AEIOU frameworks: This is a way to make observations, and stands for activities, environments, interactions, objects, and users. It serves as a series of prompts to remind the observer to record the multiple dimensions of a situation with textured focus on the user and their interactions with their environment.

ARIZ (algorithm for creative problem solving): A procedure to guide the TRIZ student from the statement of the IFR (ideal final result) to a redefinition of the problem to be solved, and then to the solutions to the problem.

Absence thinking: *Absence thinking involves training the mind to think creatively about what it is thinking and not thinking. When you are thinking about a specific something, you often notice what is not there, you watch what people are not doing, and you make lists of things that you normally forget.*

Abstract rules: Abstract rules are those unarticulated, yet essential, guidelines, norms, and traditions that people within a social setting tend to follow.

Abundance and redundancy: Abundance and redundancy are based on the belief (not necessarily factual) that if you want a good invention that solves a problem, you need lots of ideas.

Administrative process: Specifies what tasks need to be done and the order in which they should be accomplished, but does not give any, or at least very little, insight as to how those tasks should be realized.

Affinity diagram: A technique for organizing a variety of subjective data into categories based on the intuitive relationships among individual pieces of information. It is often used to find commonalties among concerns and ideas. It lets new patterns and relationships between ideas be discovered.

Affordable loss principle: Stipulates that entrepreneurs risk no more than they are willing to lose.

Agile innovation: A procedure used to create a streamlined innovation process that involves everyone. If an innovation process already exists, then the procedure can be used to improve the process, resulting in a reduction of development time, resources required, costs, delays, and faults.

Analogical thinking and mental simulations: Using past successes applied to similar problems by mental simulations and testing.

Architect: Designs (or authorizes others to design) an end-to end, integrated innovation process, and also promotes organization design for innovation, where each function contributes to innovation capability.

Attribute listing, morphological analysis, and matrix analysis: Techniques that are good for finding new combinations of products or services. We use attribute listing and morphological analysis to generate new products and services. Matrix analysis focuses on businesses. It is used to generate new approaches, using attributes such as market sectors, customer needs, products, promotional methods, etc.

Attributes-based questions: Questions based on attributes are ones in which you look for a specific attribute of an object or idea.

Balanced breakthrough model: Suggests that successful new products and services are desirable for users, viable from a business perspective, and technologically feasible.

Barrier buster: Helps navigate political landmines and removes organizational obstacles.

Benchmark (BMK): Standard by which an item can be measured or judged.

Benchmarking (BMKG): *A systematic way to identify, understand, and creatively evolve superior products, services, designs, equipment, processes, and practices to improve your organization's real performance.*

Benchmarking innovation: A form of contradiction. Doing something completely new—applying an invention in a new way—it means that others are not doing the same thing. Thus, there is nothing to benchmark.

Biomimicry: *Biomimetic or biomimicry is the imitation of the models, systems, and elements of nature for the purpose of solving complex human problems (Wikipedia). It is the transfer of ideas from biology to technology; the design and production of materials, structures, and systems that are modeled on biological entities and processes. The process involves understanding a problem and observational capability together with the capacity to synthesize different observations into a vision for solving a problem.*

Bottom-up planning for innovation: A process where innovations are described in portfolio requirements to meet business objectives.

Brainstorming: *A technique used by a group to quickly generate large lists of ideas, problems, or issues. The emphasis is on quantity of ideas, not quality.*

Brainstorming or operational creativity: Brainstorming combines a relaxed, informal approach to problem solving with lateral thinking. In most cases, brainstorming provides a free and open environment that encourages everyone to participate. While brainstorming can be effective, it is important to approach it with an open mind and a spirit of nonjudgment.

Brainwriting 6-3-5: *An organized brainstorming with writing technique to come up with ideas in the aid of innovation process stimulating creativity.*

Breakthrough, disruptive, new-to-the-world innovation: Paradigm shifts that reframe existing categories. Disruptive innovation drives significant, sustainable growth by creating new consumption occasions and transforming or obsolescing markets.

Bureaucratic process: Occurs where the inputs are defined and a specific routine is performed, but the desired output is obtained only by random chance.

Business case: *A business case captures the reasoning for initiating a project or task. It is most often presented in a well-structured written* document, but in some cases may come in the form of a short verbal agreement or presentation.

Business innovation maturity model (BIMM): Offers a road map to innovation management maturity.

Business model generation canvas: This is a tool that maps what exists. The business model canvas is a strategic management and entrepreneurial tool comprising the building blocks of a business model. The business is expressed visually on a canvas with the articulation of the nine interlocking building blocks in four cluster areas: offering—value proposition; customer—customer segments; customer relationships; infrastructure—distribution channels, key resources, key partnerships, key activities; value—cost structure and revenue model.

Business model innovation: Changes the method by which an organization creates and delivers value to its customers and how, in turn, it will generate revenue (capture value).

Business plan: A business plan is a formal statement of a set of business goals, the reason they are believed to be obtainable, and the plan for reaching these goals. It also contains background information about the organization or team attempting to reach these goals.

Capital investment: This is the cost of manufacturing equipment, packaging equipment, and change parts.

Cause-and-effect diagram: A visual representation of possible causes of a specific problem or condition. The effect is listed on the right-hand side and the causes take the shape of fish bones. This is the reason it is sometimes called a "fishbone diagram" or an "Ishikawa diagram."

Co-creation innovation: A way to introduce external catalysts, unfamiliar partners, and disruptive thinking into an organization in order to ignite innovation. The term co-creation innovation can be used in two ways: co-development and the delivery of products and services by two or more enterprises, and co-creation of products and services with customers.

CO-STAR: Specifically designed to focus the creativity of innovators on ideas that matter to customers and have relevance in the market. It is an easy-to-use tool for turning raw ideas into powerful value propositions.

Collective effectiveness: In a complex and highly competitive business environment, it is hard to sustain or support research and development (R&D) and innovation expenses. Networking allows firms access to different external resources like expertise, equipment, and overall know-how that has already been proven with less cost and in a shorter period.

Collective learning: Networking not only helps firms gain access to expensive resources like machinery, laboratory equipment, and technology, but it also facilitates shared learning via experience and good practice sharing events. This brings new insight and ideas for a firm's current and future innovation projects.

Combination methods: *A by-product of already applied processes, systems, products, and service-wise solutions integrated into a one solution system to produce one end-result that is unique.*

Communication of innovation information: Employees vary greatly in their ability to evaluate potentially significant market information and convey qualified information to pertinent receivers in the product development stream.

Comparative analysis: *A detailed study/comparison of an organization's product or service to the competitors' comparable product or service.*

Competitive analysis: *It consists of a detailed study of an organization's competitor products, services, and methods to identify their strengths and weaknesses.*

Competitive shopping (sometimes called mystery shopper): *The use of an individual or a group of individuals that go to a competitor's facilities or directly interact with the competitor's facilities to collect information related to how the competitor's processes, services, or products are interfacing with the external customer. Data are collected related to key external customer impact areas and compared with the way the organization is operating in those areas.*

Conceptual clustering: *The inherent structure of the data (concepts) that drives cluster formation. Since this technique is dependent on language, it is open to interpretation and consensus is required.*

Concept tree (see Conceptual clustering).

Confirmation bias: *The tendency of people to include or exclude data that does not fit a preconceived position.*

Consumer co-creation: *Fostering individualized interactions and experience outcomes between a consumer and the producer of the organization output. This can be done throughout the whole product life cycle. Customers may share their needs and comments, and even help spread the word or create communities in the commercialization* phase. *This approach provides a one-time limited interaction with consumers. Today, it is possible to enable constant interactions to really transfer knowledge, needs, desires, and trends from the consumer in a more structured way: co-creation.*

Contingency planning: *A process that primarily delivers a risk management strategy for a business to deal with the unexpected events effectively and the strategy for the business recovery to the normal position. The output of this process is called "contingency plan" or "business continuity and recovery plan."*

Contradiction analysis: The process of identifying and modeling contradictory requirements within a system, which, if left unresolved, will limit the performance of the system in some manner.

Contradictions: TRIZ defines two kinds of contradictions: physical and technical.

Convergent thinking: Vetting the various ideas to identify the best workable solutions.

Copyrights: Legal protection of original works of artistic authorship.

Core or line extensions, renovation, sustaining close-in innovation: Extends and adds value to an existing line or platform of products via size, flavor, or format. It is incremental improvement to existing products.

Cost–benefit analysis (CBA): *A financial analysis where the cost of providing (producing) a benefit is compared with the expected value of the benefit to the customer, stakeholder, etc.*

Counseling and mentoring: These people love teaching, coaching, and mentoring. They like to guide employees, peers, and even their clients to better performance.

Crazy quilt principle: Based on the expert entrepreneur's strategy to continuously seek out people who may become valuable contributors to his or her venture.

Create: Make something; to bring something into existence. The difference between creativity and innovation is that the output from innovation has to be a value-added output while the output from creativity does not have to be value added.

Creative (preferred definition): Using the ability to make or think of new things involving the process by which new ideas, stories, products, etc., are created.

Creative problem solving (CPS): *A methodology developed in the 1950s by Osborn and Parnes. The method calls for solving problems in* *sequential stages with the systematic alternation of divergent and convergent thinking. It can be enhanced by the use of various creative tools and techniques during different stages of the process.*

Creative production: These people love beginning projects, making something original, and making something out of nothing. This can include processes or services as well as tangible objects. They are most engaged when inventing unconventional solutions. In an innovation process, these people may thrive on the ideation phase, creating multiple solutions to the identified problems.

Creative thinking: Creative thinking is all about finding fresh and innovative solutions to problems, and identifying opportunities to improve the way that we do things, along with finding and developing new and different ideas. It can be described as a way of looking at problems or situations from a fresh perspective that suggests unorthodox solutions, which may look unsettling at first.

Creativity: Creativity is the mental ability to conceptualize or imagine new, unusual, or unique ideas, to see the new connection between seemingly random or unrelated things.

Cross-industry innovation: Refers to innovations stemming from cross-industry affinities and approaches involving transfers from one industry to another.

Crowdfunding: The collective effort of individuals who network and pool their money, usually via the Internet, to support efforts initiated by other people or organizations.

Crowdsourcing: A term for a varied group of methods that share the attribute that they all depend on some contribution from the crowd. According to Howe, it is the act of a company or institution taking a function once performed by employees and outsourcing it to an undefined (and generally large) network of people in the form of an *open call.*

Culture: Culture is all about how people behave, treat each other, and treat customers.

Culture creator: Ensures the spirit of innovation is understood, celebrated, and aligned with the strategy of the organization.

Customer advocate: Keeps the voice of the customer alive in the hearts, minds, and actions of innovators and teams.

Customer profile: Empathy map is a technique for creating a profile of your customer beyond the simple demographics of age, gender, and income that has been in use for some time.

DVF model (desirable, viable, feasible): Another name for the balanced breakthrough model.

Design innovation: Focuses on the functional dimension of the job to be done, as well as the social and emotional dimensions, which are sometimes more important than functional aspects.

Design for X (DFX): Both a philosophy and methodology that can help organizations change the way that they manage product development and become more competitive. DFX is defined as a knowledge-based approach for designing products to have as many desirable characteristics as possible. The desirable characteristics include quality, reliability, serviceability, safety, user-friendliness, etc. This approach goes beyond the traditional quality aspects of function, features, and appearance of the item.

Design of experiments: This method is a statistically based method that can reduce the number of experiments needed to establish a mathematical relationship between a dependent variable and independent variables in a system.

Directed innovation: Directed innovation is a systematic approach that helps cross-functional teams apply problem-solving methods like brainstorming, TRIZ, creative problem solving, Six Thinking Hats™, Lateral Thinking™, assumption storming, inventing, Question Banking™, and provocation to a specifically defined problem in order to create novel and patentable solutions.

Direction setter: Creates and communicates vision and business strategy in a compelling manner, and ensures innovation priorities are clear.

Disruptive innovation: A process where a product or service takes root initially in simple applications at the bottom of a market and then relentlessly moves upmarket, eventually displacing established competitors.

Divergent thinking: Coming up with many ideas or solutions to a problem.

Diversity trumps ability theorem: States that a randomly selected collection of problem solvers outperforms a collection of the best individual problem solvers.

Drive to acquire: The drive to acquire tangible goods such as food, clothing, housing, and money, but also intangible goods such as experiences, or events that improve social status.

Drive to bond: The need for common kinship bonding to larger collectives such as organizations, associations, and nations.

Drive to comprehend: People want to be challenged by their jobs, to grow and learn.

Drive to defend: Includes defending your role and accomplishments. Fulfilling the drive to defend leads to feelings of security and confidence.

Edison method: Edison method consists of five strategies that cover the full spectrum of innovation necessary for success.

Effectuation: Taking action toward unpredictable future states using currently controlled resources and with imperfect knowledge about current circumstances.

Ekvall: Ekvall's model of the creative climate identifies 10 factors that need to be present:

- Idea time
- Challenge
- Freedom
- Idea support
- Conflicts
- Debates
- Playfulness, humor
- Trust, openness
- Dynamism
- Liveliness

Elevator speech: An elevator speech is a clear, brief message or "commercial" about the innovative idea you are in the process of implementing. It communicates what it is, what you are looking for, and how it can benefit a company or organization. It is typically no more than two minutes, or the time it takes people to ride from the top to the bottom of a building in an elevator.

Emergent collaboration: A social network activity where a shared perspective emerges from a group through spontaneous (unplanned) interactions.

Emotional rollercoaster: It is a notion, similar to journey mapping, that identifies areas of high anxiety in a process and, as such, exposes opportunities for new solutions.

Enterprise control: These people love to run projects or teams and control the assets. They enjoy owning a transaction or sale, and tend to ask for as much responsibility as possible in work situations.

Entrepreneur: Someone who exercises initiative by organizing a venture to take benefit of an opportunity and, as the decision maker, decides what, how, and how much of a good or service will be produced. An entrepreneur supplies risk capital as a risk taker, and monitors and controls the business activities. The entrepreneur is usually a sole proprietor, a partner, or the one who owns the majority of shares in an incorporated venture. From the businessdictionary.com.

Era-based questions: Era-based questions require that you put yourself in the position of thinking about a question in a different time or place from the one you are currently in.

Experiments: In this context, experiments represent a mixture of surveys and observations in an artificial setting and can be summarized as test procedures.

Ethnography: Ethnography can be used in many ways, but most significantly in the creation of a new product or service with a clear understanding of the many different ways that a person may accomplish a task based on their own world view. Observing and recording what people do to solve a problem and not what they say the problems are. It is based on anthropology but used on current human activities. It is based on the belief that what people do can be more reliable than what they say.

FAST: An innovative technique to develop a graphical representation showing the logical relationships between the functions of a project, product, process, or service based on the questions "How" and "Why." In this case, it should not be confused with FAST, which stands for the Fast Action Solution Team methodology created by H.J. Harrington. The two are very different in application and usage.

Failure mode effects analysis: A matrix-based method used to investigate potential serious problems in a proposed system before the final design. It creates a risk priority number that can be used to create a ranking of the biggest risks and then ranks the proposed solution.

Financial management: Activities and management of financial programs and operations, including accounting liaison and pay services; budget preparation and execution; program, cost, and economic analysis; and nonappropriated fund oversight. It is held responsible and accountable for the ethical and intelligent use of investors' resources.

Financial reporting: Includes the main financial statements (income statement, balance sheet, statement of cash flows, statement of retained earnings, statement of stockholders' equity) plus other financial information such as annual reports, press releases, etc.

Fishbone diagrams, also known as Ishikawa diagrams: A mnemonic diagram that looks like the skeleton of a fish and has words for the major spurs that prompt causes for the problem.

Five dimensions of a service innovation model:
- Organizational
- Product
- Market
- Process
- Input

Flowcharting: A method of graphically describing an existing or proposed process by using simple symbols, lines, and words to pictorially display the sequence of activities. Flowcharts are used to understand, analyze, and communicate the activities that make up major processes throughout an organization. They can be used to graphically display movement of product, communications, and knowledge related to anything takes an input and value to it and produces an output.

Focus group: Made up of a group of individuals that are knowledgeable or would make use of the subject being discussed. The facilitator is used to lead the discussions and record key information related to the discussions.

Focus group: A focus group is a structured group interview of typically 7 to 10 individuals who are brought together to discuss their views related to a specific business issue. The group is brought together so that the organizer can gain information and insight into a specific subject or the reaction to a proposed product.

Force field analysis: A visual aid for pinpointing and analyzing elements that resist change (restraining forces) or push for change (driving forces). This technique helps drive improvement by developing plans to overcome the restrainers and make maximum use of the driving forces.

Four dimensions of innovation:
- Technology: technical uncertainty of innovation projects
- Market: targeting of innovations on new or not previously satisfied customer needs
- Organization: the extent of organizational change
- Innovation environment: impact of innovations on the innovation environment

Four-square model: The four-square model is a design process that consists of five steps:

- Problem framing: Identify what problem we intend to solve and outline a general approach for how we will solve it.
- Research: Gather qualitative and quantitative data related to the problem frame.
- Analysis: Unpack and interpret the data, building conceptual models that help explain what we found.
- Synthesis: Generate ideas and recommendations using the conceptual model as a guide.
- Decision making: Conduct evaluative research to determine which concepts or recommendations best fit the desirable, viable, and feasible criteria.

Four-square model for design innovation: Composed of two sets of polar extremes: understand-make and abstract-concrete.

Functional analysis: A standard method of systems engineering that has been adapted into TRIZ. The subject–action–object method is most frequently used now. It is a graphical and primarily qualitative methodology used to focus the problem solver on the functional relationships (good or bad) between system components.

Functional model: A structured representation of the functions (activities, actions, processes, operations) within the modeled system or subject area.

Functional innovation: Involves identifying the functional components of a problem or challenge, and then addressing the processes underlying those functions that are in need of improvement. Through this process, overlaps, gaps, discontinuities, and other inefficiencies can be identified.

Futurist: Looks toward the future, scouts new opportunities, helps everyone see their potential. Enables people throughout the organization to discover the emerging trends that most influence their work.

Generic creativity tools: Generic creativity tools are a set of commonly used tools that are designed to assist individuals and groups to originate new and different thought patterns. They have many common characteristics like thinking positive, not criticizing ideas, thinking out of the box, right-brain thinking, etc. Some of the typical tools are benchmarking, brainstorming, Six Thinking Hats, storyboarding, and TRIZ.

Goal: The end toward which effort is directed: the terminal point in a race. These are always specified in time and magnitude so they are easy to measure. Goals have key ingredients. First, they specifically state the target for the future state and, second, they give the time interval in which the future state will be accomplished. These are key input to every strategic plan.

Goal-based questions: Goal-based questions pose the end goal without specifying the means or locking you into particular attributes.

Go-to-market investment: This is the cost of slotting fees for distribution, trade spending, advertising dollars (creative development, media spend), promotional programs, and digital/social media.

Gupta's Einsteinian theory of innovation (GETI): This theory states: Thus, every idea must have some energy associated with it that is an outcome of effort and the speed of the thought. Expressed as

- Innovation value = resources $\times$ (speed of thought)^2

where the speed of thought can be described by the following relationship:

- Speed of thought $\equiv$ function (knowledge, play, imagination)

HU Diagrams: Effective ways of providing a visual picture of the interface between the harmful and useful characteristics of a system or process.

Hitchhiking: When a breakthrough occurs, it is a fertile area for innovators. One should hitchhike on the breakthrough to create new applications and improvements that can be inventions.

I-TRIZ: An abbreviation for Ideation TRIZ, which is a restructuring and enhancement of classical TRIZ methodology based on modern research and practices. It is a guided set of step-by-step questions and instructions that aid teams in approaching, thinking, and dealing with systems targeted for innovation. It provides specific practical team guidance for the following applications:

- *Solving a nontechnical or business issue*
- *Solving a technical or engineering issue*
- *Finding the root cause(s) of a system issue*
- *Anticipating and preventing possible systems*
- *Predicting and inventing specific innovative products or services customers will want in the future*
- *Patent (invention) evaluation, preparation, and enhancement to either work-around (invent around, design around) an existing (blocking) patent or provide a patent "fence" to prevent possible work-around*

Ideal final result (IFR): This states that in order to improve a system or process, the output of that system must improve (i.e., volume, quantity, quality, etc.), the cost of the system must be reduced, or both. It is an implementation-free description of the situation after the problem has been solved.

Idea priority index: Prioritizes ideas based on the potential cost–benefit analysis, associated risks, and likely time to commercialize the idea.

Idea selection by grouping or tiers: Groups can be helpful in evaluating tiers like top ideas or worst ideas. Both grouping and tiers are only useful in a batch evaluation process, not a continuous process.

Idea selection by checklist or threshold: An individual idea's list of attributes must match the preset checklist or threshold in order to pass (e.g., be implemented in six months, profit at least $500,000, and require no more than two employees).

Idea selection by personal preference: A manager, director, line employee, or even expert is used to screen an idea based on his or her own preferences.

Idea selection by point scoring: Uses a scoring sheet to rate a particular idea on its attributes (e.g., an idea that can be implemented in six months gets +5 points, and one that can make more than × dollars gets +10 points). The points are then added together, and the top ideas are ranked by highest total point scores.

Idea selection by priority index (IPI): The IPI prioritizes ideas on the basis of the potential cost–benefit analysis, associated risks, and likely time to commercialize the idea, using the following relationships:

- Annualized potential impact of the idea = ($) × probability of acceptance
- IPI = annualized cost of idea development ($) × time to commercialize (year)

Idea selection by voting: Individual(s) can vote openly or in a closed ballot (i.e., blind or peer review). Voting can be weighted or an individual, such as expert, can give multiple votes to a given idea.

Image board, storyboarding, role playing: These are collections of physical manifestations (image collages or product libraries) of the desirable (or undesirable if you are using that as a motivator) to help generate ideas, or to facilitate conversations with users about what they want.

Imaginary brainstorming: It expands the brainstorming concept past the small group problem-solving tool to an electronic system that presents the problem/opportunity to anyone that is approved to participate in the electronic system. Creative ideas are collected, and a smaller group is used to analyze and identify innovative, imaginative concepts.

Indexing: Providing a tag for a fact, piece of information, or experience, so that you can retrieve it when you want it.

Influence through language and ideas: These people love expressing ideas for the enjoyment of storytelling, negotiating, or persuading. This can be in written or verbal form, or both.

Innovation: An advancement that transcends a limiting situation within the system under analysis. Another way to describe these limiting situations is to refer to them as contradictory requirements within a system.

Innovation: Converting ideas into tangible products, services, or processes. The challenge that every organization faces is how to convert good ideas into profit. That is what the innovation process is all about.

Innovation (preferred definition): The process of translating an idea or invention into an intangible product, service, or process that creates value for which the consumer (the entity that uses the output from the idea) is willing to pay for it more than the cost to produce it.

Innovation benchmarking: Comparing one organization, process, or product to another that is considered a standard.

Innovation blueprint: A visual map to the future that enables people within an enterprise or community to understand where they are headed and how they can build that future together. The blueprint is not a tool for individual innovators or teams to improve a specific product or service, or to create new ones. Rather, the innovation blueprint is a tool for designing an enterprise that innovates extremely effectively on an ongoing basis.

Innovation culture: A culture that requires continuous learning, practices, and exceptions of risk and failure; holds individuals accountable for action; and has aggressive timing.

Innovation management: The collection of ideas for new or improved products and services, and their development, implementation, and exploitation in the market.

Innovation master plan framework: *The innovation master plan framework consists of five major elements: strategy, portfolio, processes, culture, and infrastructure.*

Innovation metrics: Measurements to validate that the organization innovating. They typically are

- Annual R&D budget as a percentage of annual sales
- Number of patents filed in the past year
- Total R&D headcount or budget as a percentage of sales
- Number of active projects
- Number of ideas submitted by employees
- Percentage of sales from products introduced in the past × year(s)

Innovation process: The innovation process is made up of seven phases:

- Phase I: Creation phase
- Phase II: Value proposition phase
- Phase III: Resourcing phase
- Phase IV: Documentation phase
- Phase V: Production phase
- Phase VI: Sales/delivery phase
- Phase VII: Performance analysis phase

Innovative categories: Most service innovations can be categorized into one of the following groups:

- Incremental or radical, based on the degree of new knowledge
- Continuous or discontinuous, depending on its degree of price performance improvements over existing technologies. Sometimes called evolutionary innovation
- Sustaining or disruptive, relative to the performance of the existing products
- Exploitative or evolutionary, innovation in terms of pursuing new knowledge and developing new services for emerging markets

Innovative problem solving: A subset of problem solving in that a solution must resolve a limitation in the system under analysis in order to be an innovative solution.

Innovator: An innovator is an individual that creates a unique idea that is marketable and guides it through the innovative process so that is value to the customer is greater than the resources required to produce.

Insight: A linking or connection between ideas in the mind. The connections matter more than the pieces.

Inspiration: The word inspiration is from the Latin word *inspire*, meaning *to blow into.*

Inspire innovation tools: They are tools that stimulate the unique creative powers. Some of them are

- Absence thinking
- Biomimicry
- Concept tree
- Creative thinking
- Ethnography
- HU diagrams
- Imaginary brainstorming
- I-TRIZ
- Mind mapping
- Open innovation
- Storyboarding
- TRIZ

Integrated innovation system: Covers the full end-to-end innovation process, and ensures the practices and tools are aligned and flow easily from one to the other.

Intellectual property rights: The expression "intellectual property rights" refers to a number of legal rights that serve to protect various products of the intellect (i.e., "innovations"). These rights, while different from one another, can and do sometimes offer overlapping legal protection.

Intersection of different sets of knowledge: Networking creates different relationships to be built across knowledge frontiers and opens up the participating organizations to new stimuli and experience.

Intrapreneur: An intrapreneur is an employee of a large corporation who is given freedom and financial support to create new products, services, systems, etc., and does not have to follow the corporation's usual routines or protocols.

Joint risk taking: Since innovation is a highly risky activity, it is very difficult for a single firm to undertake it by itself and this impedes the development of new technologies. Joint collaboration minimizes the risk for each firm and encourages them to engage in new activities. This is the logic behind many precompetitive consortia collaborations for risky R&D.

Journey map or experience evaluations: It is a diagram that illustrates the steps your customer(s) go through in engaging with your company, whether it is a product, an online experience, a retail experience, a service, or any combination of these.

Kano analysis: *A pictorial way to look at customer levels of dissatisfaction and satisfaction to define how they relate to the different product characteristics. The Kano method is based on the idea that features can be plotted using axes of fulfillment and delight. This defines areas of must haves, more is better, and delighters. It classifies customer preferences into five categories:*

- Attractive
- One-dimensional
- Must-be
- Indifferent
- Reverse

Kepner Trego: This method is very useful for processes that were performing well and then developed a problem. It is a good step-by-step method that is based on finding the cause of the problem by asking what changed since the process was working fine.

Knowledge management (KM): *A strategy that turns an organization's intellectual assets, both recorded information and the talents of its members, into greater productivity, new value, and increased competitiveness. It is the leveraging of collective wisdom to increase responsiveness and innovation.*

Leadership metrics: Leadership metrics address the behaviors that senior managers and leaders must exhibit to support a culture of innovation.

Lemonade principle: Based on the old adage that goes, "If life throws you lemons, make lemonade." In other words, make the best of the unexpected.

Link between climate and organizational innovation: Nine areas need to be evaluated to determine this linkage:

- Challenge, motivation
- Freedom
- Trust
- Idea time
- Play and humor
- Conflicts
- Idea support
- Debates
- Risk taking

Live-ins, shadowing, and immersion labs: They are designed to resemble the retail or home environment and gather extensive information about product purchase or use. These laboratories are used to both test the known, launch new products, and observe user behavior.

Lotus blossom: This technique is based on the use of analytical capacities and helps generate a great number of ideas that will possibly provide the best solution to the problem to be addressed by the management group. It uses a six-step process.

Managing people and relationships: Unlike counseling and mentoring people, these people live to manage others on a day-to-day basis.

Marketing research: Can be defined as the systematic and objective identification, collection, analysis dissemination, and use of information that is undertaken to improve decision making related to products and services that are provided to external customers.

Market research tools: The following are typical marketing research tools:

- Analysis of customer complaints
- Brainstorming
- Contextual inquiry, empathic design
- Cross-industry innovation
- Crowdsourcing
- In-depth interview
- Lead user technique
- Listening-in technique
- Netnography
- Outcome-driven innovation
- Quality function deployment
- Sequence-oriented problem identification, sequential incident technique
- Tracking, panel
- Analytic hierarchy process
- Category appraisal
- Concept test, virtual concept test
- Conjoint analysis
- Store and market test
- Free elicitation
- Information acceleration
- Information pump
- Kelly repertory grid
- Laddering
- Perceptual mapping

- Product test, product clinic
- Virtual stock market, securities trading of concepts
  i. Zaltman metaphor elicitation technique
  ii. Customer idealized design
  iii. Co-development
  iv. Expert Delphi discussion
  v. Focus group
  vi. Future workshop
  vii. Tool kit

Matrix diagram (decision matrix): *A systematic way of selecting from larger lists of alternatives. They can be used to select a problem from a list of potential problems, select primary root causes from a larger list, or to select a solution from a list of alternatives.*

Medici effect: The book by this name describes the intersection of significantly different ideas that can produce cross-pollination of fields and create more breakthroughs.

Mentor: Coaches and guides innovation, champions, and teams.

Methodology merger: Each methodology brings with it certain strengths and weaknesses that serve to fulfill specific steps and activities represented on the problem-solving pathway. When combined together and properly utilized, these methodologies create a very effective and useful outcome.

Mind mapping: *An innovation tool and method that starts with a main idea or goal in the middle, and then flows or diagrams ideas out from this one main subject. By using mind maps, you can quickly identify and understand the structure of a subject. You can see the way that pieces of information fit together in a format that your mind finds easy to recall and quick to review. They are also called spider diagrams.*

Mini problem: One that is solved without introducing new elements. We have to understand resources, since the emphasis is on solving the problem without introducing anything new to the system.

Moccasins/walking in the customer's shoes: The moccasins approach is more often called walking in the customer's shoes. This activity allows members of the organizations group to directly participate in the process that the potential customer is subjected to by physically playing the role of the customer.

Myers–Briggs (MBTI): This is a survey-style measurement instrument used in determining an individual's social style preference.

NSD: NSD is an abbreviation for new service development.

Network-centric approach: Network-centric approach is taught in colleges and based on collaborative brainstorming. The concept is that more minds are better than one at a given time.

Networker: Works across organizational boundaries to engage stakeholders, promotes connections across boundaries, and secures widespread support.

Nominal group technique: A technique for prioritizing a list of problems, ideas, or issues that gives everyone in the group or team equal voice in the priority setting process.

Nonalgorithmic interactions: Actions with cognitive and physical materials of a project whose results you cannot predict for certain, those results you do not know.

Nonprobability sampling techniques: Use samples drawn according to specific and considered characteristics and are therefore based on the researcher's subjective judgment.

Nonprofit: An organization specifically formed to provide a service or product on a not-for-profit basis as determined by applicable law.

Observation: In this context, observation means the recording of behavioral patterns of people, objects, and events in order to obtain information.

Online collaboration: Convening an online brainstorming or idea generation session so members can participate remotely, instead of organizing a group in a room together.

Online management platforms: These are used to foster innovation enable large groups of people to innovate together—across geographies and time zones. Users can post ideas and value propositions online and can collaborate with others to make these stronger. The community can rate and rank ideas or value propositions, post comments and recommendations, link to resources, build on each other's ideas, and support each other to improve each other's innovations.

Open innovation: The use and application of collective intelligence to produce a creative solution to a challenging problem, as well as to organize large amounts of data and information. The term refers to the use of both inflows and outflows of knowledge to improve internal innovation and expand the markets for external exploitation of innovation. The central idea behind open innovation is that, in a world of widely distributed knowledge, companies cannot afford to rely entirely on their own research, but should instead buy or license processes or inventions (i.e., patents) from other companies.

Opportunity-driven model: An opportunity-driven model is more representative of street-smart individuals who take an idea at the right time and the right place, devise a solution, know how to market it, and capitalize on their breakthrough. They also appear to be lucky.

Organizational capability metrics: Organizational capability metrics focus on the infrastructure and process of innovation. Capability measures provide focus for initiatives geared toward building repeatable and sustainable approaches to invention and reinvention.

Organizational change management (OCM): *A comprehensive set of structured procedures for the decision making, planning, executing, and evaluation activities. It is designed to minimize the resistance and cycle time to implementing a change.*

Organizational effectiveness measurements: The following is a typical way of measuring the organizations innovation effectiveness. Typically, it is measured in four key areas of management processes: product, sales, internal services, and sales and marketing. Each area is typically evaluated in a combination of the following:
- Committed leadership
- Clear strategy
- Market insights
- Creative people
- Innovative culture
- Competitive technologies
- Effective processes
- Supportive infrastructure
- Managed projects

Organization internal boundaries: Employee silos often isolate chains of command and communication, which can impede the progress of a valuable idea through product development.

Osborn method: Original brainstorming method developed by Alex F. Osborn by primarily requiring solicitation of unevaluated ideas (divergent thinking), followed by convergent organization and evaluation.

Outcome-driven innovation (ODI): *Built around the theory that people buy products and services to complete tasks or jobs they value. As people complete these jobs, they have certain measurable outcomes that they are attempting to achieve. It links a company's value creation activities to customer-defined metrics. Included in this method is the opportunity algorithm, which helps designers determine the needs that satisfied customers has. This helps determine which features are most important to work on. Most important is this tool's intention of trying to find unmet needs that may lead to new and innovative products/services.*

PESTEL frameworks: The PESTEL framework focuses on the macroeconomic factors that influence a business. These factors are

- Political factors
- Economic factors
- Social factors
- Technological factors
- Environmental factors

Patents: A government-granted right that literally and strictly permits the patent owner to prevent others from practicing the claimed invention.

Performance engine project: A project that seeks to improve a current level of performance and not to create a new value proposition.

Permeability to innovation idea sources: Information and idea seeking differs greatly among companies.

Physical contradictions: Situations where one requirement has contradictory, opposite values to another.

Pilot in the plane principle: Based on the concept of *control*, using effectual logic and is referred to as *nonpredictive control*. Expert entrepreneurs believe they can determine their individual futures best by applying effectual logic to the resources they currently control.

Pipeline model: A pipeline model as driven by chance or innate genius, is a somewhat common perception of the innovation process. Inventors who work in research drive the pipeline model and development environment on a specific topic, explore new ideas, and develop new products and services.

Plan–do–check–act (PDCA): *A structured approach for the improvement of services, products, or processes. It is also sometimes referred to as plan–do–check–adjust. Another version of this PDCA cycle is OPDCA. The added "O" stands for observation or, as some versions say, "Grasp the current condition."*

Platform: A consumer need–based opportunity that inspires multiple innovation ideas with a sustainable competitive advantage to drive growth.

Portfolio management: The ongoing management of innovation to ensure delivery against stated goals and innovation strategy.

Post-Fordist: Companies after the Henry Ford efficient production era where managers wielded inordinate responsibility for profit and loss and the new postmodern leaders of the global economy, who are responsible for developing talented teams.

Potential investor presentation: A short PowerPoint presentation designed to convince an individual or group to invest their money in an organization or a potential project. It can be a presentation to an individual or group not part of the organization or the management of the organization that the presenter is presently employed by. It is usually part of a short meeting that usually lasts no more than one hour.

Practices: To look at all the inputs that we have available for selection and all the available operations or routines that we can perform on those inputs, then to select those inputs and operations that will give us our desired results.

Primary data: Data collected from the field or expected customer.

Primary sources: Gathered directly from the source; for instance, if new customer opinions were required to justify a new product, then customer interviews, focus groups, or surveys would suffice.

Principles of invention: A set of 40 principles from a variety of fields such as software, health care, electronics, mechanics, ergonomics, finance, marketing, etc., used to solve problems.

Proactive personal creativity: Proactive strategies to be especially effective in increasing the originality and effectiveness of personal creativity:

- Self-trust
- Open up
- Clean and organize
- Make mistakes
- Get angry
- Get enthusiastic
- Listen to hunches
- Subtract instead of adding
- Physical motion
- Question the questions
- Pump up the volume
- Read, read, read

Probability sampling techniques: Use samples randomly drawn from the whole population.

Probe-and-learn strategy: Where nonworking prototypes are developed in rapid succession, tested with potential customers, and feedback is sought on each prototype.

Problem detection and affinity diagrams: Focus groups, mall intercepts, or mail and phone surveys that ask customers what problems they have. They are all forms of problem detection. The responses are grouped according to commonality (affinity diagrams) to strengthen the validity of the response. Developing the correct queries and interpreting the responses are critical to the usefulness of the method.

Problem solving: Generating a workable solution.

Process: *A series of interrelated activities or tasks that take an input and produces an output.*

Process phases:

> Phase 1: Opportunity identification
> Phase 2: Idea generation
> Phase 3: Concept evaluation
> Phase 4: Acquiring resources
> Phase 5: Development
> Phase 6: Producing the product
> Phase 7: Launch
> Phase 8: Sales and marketing
> Phase 9: Evaluation of results

Process innovation: Innovation of internal processes. New or improved delivery methods may occur in all aspects of the supply chain in an organization.

Process redesign: *A methodology used to streamline a current process with the objective of reducing cost and cycle time by 30% to 60% while improving output quality from 20% to 200%.*

Process reengineering: *A methodology used to radically change the way a process is designed by developing an aggressive vision of how it should perform and using a group of enablers to prepare a new process design that is not hampered by the present processes paradigms. Use when a 60% to 80% reduction in cost or cycle time is required. Process reengineering is sometimes referred to as new process design or process innovation.*

Product innovation: A multidisciplinary process usually involving many different functions within an organization and, in large organizations, often in coordination across continents.

Project management: The application of knowledge, skills, tools, and techniques to project activities in order to meet or exceed stakeholders' needs and expectations from a project. (Source: PMBOK Guide.)

Proof of concept (POC): A demonstration, the purpose of which is to verify that certain concepts or theories have the potential for real-world application. A proof of concept is a test of an idea made by building a prototype of the application. It is an innovative, scaled-down version of the system you intend to develop. The proof of concept provides evidence that demonstrates that a business model, product, service, system, or idea is feasible and will function as intended.

Pyramiding: A search technique in which the searcher simply asks an individual (the starting point) to identify one or more others who he or she thinks has higher levels of expertise.

Qualitative research: Gathered data are transcribed, and single cases are analyzed and compared in order to find similarities and differences to gain deeper insights into the subject of interest. In *quantitative research*, the data preparation step contains the editing, coding, and transcribing of collected data.

Qualitative research (survey): Represents an unstructured, exploratory research methodology that makes use of psychological methods and relies on small samples, which are mostly not representative.

Quality function deployment (QFD), also known as the house of quality: This creates a matrix that looks like a house that can mediate the specifications of a product or process. There are subsequent derivative houses that further mediate downstream implementation issues.

Quantitative analysis: These people love to use data and numbers to figure out business solutions. They may be in classic quantitative data jobs, but may also like building computer models to solve other types of business problems. These people can fall into two camps: (a) descriptive and (b) prescriptive.

Quantitative research (survey): Can be seen as a structured research methodology based on large samples. The main objective in quantitative research is to quantify the data and generalize the results from the sample, using statistical analysis methods.

Quickscore creativity test: *A 3-minute test that helps assess and develop business creativity skills.*

ROI metrics: ROI metrics address two measures: resource investments and financial returns. ROI metrics give innovation management fiscal discipline and help justify and recognize the value of strategic initiatives, programs, and the overall investment in innovation.

Radical innovation: A high level of activity in all four dimensions, while incremental innovations (low degree of novelty) are only weakly to moderately developed in the four dimensions.

Ranking or forced ranking: Ideas are ranked (#1, 2, 3, etc.)—this makes the group consider minor differences in ideas and their characteristics. For forced ranking, there can only be a single #1 idea, a single #2 idea, and so on.

Rating scales: An idea is rated on a number of preset scales (e.g., an idea can be rated on a 1 to 10 on implementation time; any idea that reaches a 9 or 10 is automatically accepted).

Reverse engineering: *This is a process where organizations buy competitive products to better understand how the competitor is packaging, delivering, and selling their product. Once the product is delivered, it is tested, disassembled, and analyzed to determine its performance, how it is assembled, and to estimate its reliability. It is also used to provide the organization with information about the suppliers that the competitors are using.*

Robust design: *Robust design is more than a tool; it is complete methodology that can be used in the design of systems (products or processes) to ensure that they perform consistently in the hands of the customer. It comprises a process and tool kit that allows the designer to assess the impact of variation that the system is likely to experience in use, and if necessary redesign the system if it is found to be sensitive.*

Role model: Provides a living example of innovation through attention and language, as well as through personal choices and actions. Key stakeholders often test the leader's words, to see if these are real. For innovation to move forward, the leader must pass these inevitable tests—to show that, yes, he or she is absolutely committed to innovation as essential to success.

Root cause analysis (RCA): A graphical and textual technique used to understand complex systems and the dependent and independent fundamental contributors, or root causes, of the issue or problem under analysis. This is a technical process in that it provides specific direction as to how to execute the method.

Rote practice: Those activities where it looks like people are engaged in finding the right routines and inputs to obtain the desired result, but are just going through the motions.

S-curve: *A mathematical model also known as the logistic curve, which describes the growth of one variable in terms of another variable over time. S-curves are found in many fields of innovation, from biology and physics to business and technology.*

SCAMPER: *A tool that helps people generate ideas for new products and services by encouraging them to think about how you could improve existing ones by using each of the seven words that SCAMPER stands for and applying it to the new product or service in order to generate additional new ideas. SCAMPER is a mnemonic that stands for*

- Substitute
- Combine
- Adapt
- Modify
- Put to another use
- Eliminate
- Reverse

SIPOC: An acronym for *supplier, input, process, output,* and *customer* model.

Scarcity of innovation opportunities: Markets have matured into commoditized exchanges.

Scenario analysis: *A process of analyzing possible future events by considering alternative possible outcomes (sometimes called "alternative worlds"). Thus, the scenario analysis, which is a main method of projections, does not try to show one exact picture of the future. Instead, the scenario analysis is used as a decision-making tool in the strategic planning process in order to provide flexibility for long-term outcomes.*

Scientific method: The scientific method is a classical method that uses a hypothesis based on initial observations and validation through testing and revision if needed.

Secondary data: Data collected through in-house (desk research).

Secondary data sources: Involve evidence gathered from someone other than the primary source of the information. Most media outlets, magazines, books, articles, trade journals, market research reports, or publisher-based information are considered secondary sources of evidence.

Service innovation: Not substantially different from product innovation in that the goal is to satisfy customers' jobs to be done, wow and retain customers, and ultimately optimize profit.

Seven key barriers to personal creativity: The seven key barriers to personal creativity are

- Perceived definitions of creativity
- Presumed uses for creativity
- Overdependence on knowledge
- Experiences and expertise
- Habits
- Personal and professional relationship networks
- Fear of failure

Simulation: *As used in innovation, simulation is the representation of the behavior or characteristics of one system through the use of another system, especially a computer program designed for the purpose. As such, it is both a strategy and a category of tools—and is often coupled with CAI (computer-aided innovation), which is an emerging simulation domain in the array of computer-aided technologies. CAI has been growing as a response to greater industry innovation demands for more reliability in new products.*

Six Sigma: A method designed for the reduction of variation in processes. The general steps used within the DMAIC (define, measure, analyze, improve, and control) and DMADV (define, measure, analyze, develop, and verify) methodologies are mostly administrative in nature. Combining Six Sigma with other tool sets pushes the process strongly toward the technical end of the scale.

Six Thinking Hats: *It is used to look at decisions from a number of important perspectives. This forces you to move outside your habitual thinking style, and helps you to get a more rounded view of a situation. The thinking is that if you look at a problem with the "Six Thinking Hats" technique, then you will solve it using any and all approaches. Your decisions and plans will mix ambition, skill in execution, public sensitivity, creativity, and good contingency planning.*

Social business: The practice of using social technologies to transform business.

Social innovation: Social innovation relates to creative ideas designed to address societal challenges—cultural, economic, and environmental issues—that are no longer simply a local or national problem but affect the well-being of the planet's inhabitants and, ultimately, corporate profits and sustainability.

Social media: Refers to using social technologies as media in order to influence large audiences.

Social networks: Networks of friends, colleagues, and other personal contacts: strong social networks can encourage healthy behaviors. They are often computers or an online community of people with a common interest who use a website or other technologies to communicate with each other and share information, resources, etc. A business-oriented social network is a website or online service that facilitates this communication.

Spontaneous order: A term that Hayek uses to describe what he calls the open society. It is created by unleashing human creativity generally in a way not planned by anyone, and, importantly, could not have been.

Stage gate process: First introduced by R.G. Cooper in 1986 in his book *Winning at New Products.*

Stakeholder: A "stakeholder" of an organization or enterprise is someone who potentially or really influences that organization, and who is likely to be affected by that organization's activities and processes. Or, even more significantly, perceives that they will be affected (usually negatively).

Statistical analysis: A collection, examination, summarization, manipulation, and interpretation of quantitative data to discover its underlying causes, patterns, relationships, and trends.

Storyboarding: Physically structures the output into a logical arrangement. The ideas, observations, or solutions may be grouped visually according to shared characteristics, dependencies on one another, or similar means. These groupings show relationships between ideas and provide a starting point for action plans and implementation sequences.

Substantial platform, transformational, adjacencies innovation: Innovations that deliver a unique or new benefit or usage occasion, within an existing or adjacent category.

Synectics: It combines a structured approach to creativity with the free-wheeling problem-solving approach used in techniques like brainstorming. It is a useful technique when simpler creativity techniques like SCAMPER, brainstorming, and random input have failed to

> generate useful ideas. It uses many different triggers and stimuli to jolt people out of established mind-sets and into more creative ways of thinking.

Systematic innovation stages: Systematic innovation can be viewed as occurring in stages

- Concept stage
- Feasibility stage
- Development stage
- Execution stage, preparation for production
- Production stage
- Sustainability stage

Systematic innovation tools:

- Analogical thinking and mental simulations
- Theory of inventive problem solving (TRIZ)
- Scientific method
- Edison method
- Brainstorming
  - i. Osborn method
  - ii. Six Thinking Hats
  - iii. Problem detection and affinity diagrams
  - iv. Explore unusual results
  - v. Ethnography
  - vi. Function analysis and fast diagrams
  - vii. Kano method
  - viii. Abundance and redundancy
  - ix. Hitchhiking
  - x. Kepner Trego
  - xi. Quality function deployment (QFD), also known as the house of quality
  - xii. Design of experiments
  - xiii. Failure mode effects analysis
  - xiv. Fishbone diagrams, also known as Ishikawa diagrams
  - xv. Five whys
  - xvi. Medici effect
  - xvii. Technology mapping and recombination
- Trial and error

System operator (also called *9 windows* or *multiscreen* method): A visual technique that is used frequently in the initial stages of TRIZ as part of problem definition.

System operator: The construction of a 3 × 3 matrix, with the rows labeled as the system, subsystem, supersystem; and the columns labeled past, present, and future.

Systems engineering: These methods are more technical than administrative processes, as they are fairly specific as to how to create and utilize the various systems engineering models. However, these methods guide the problem solver, understanding that full system analysis is necessary in creating truly effective solutions. Therefore, these methods may be more administrative in nature than technical.

Systems engineering, system analysis: A technique to ensure that full system effects, impacts, benefits, and responses are understood when looking at changes or problems within a system.

Systems thinking: An approach to problem solving, by viewing "problems" as parts of an overall system, rather than reacting to specific part, outcomes, or events and potentially contributing to further development of unintended consequences.

TEDOC methodology: TEDOC stands for target, explore, develop, optimize, and commercialize data to solve problems creatively. As such, TRIZ brings repeatability, predictability, and reliability to the problem-solving process with its structured and algorithmic approach.

TRIZ (pronounced "treesz"): A Russian acronym for "teoriya resheniya izobretatelskikh zadatch," the theory of inventive problem solving, originated by Genrich Altshuller in 1946. It is a broad title representing methodologies, tool sets, knowledge bases, and model-based technologies for generating innovative ideas and solutions. It aims to create an algorithm to the innovation of new systems and the refinement of existing systems, and is based on the study of patents of technological evolution of systems, scientific theory, organizations, and works of art.

Technical contradictions: Classical engineering and management trade-offs. The desired state cannot be reached because something else in the system prevents it. The TRIZ patent research classified 39 features for technical contradictions.

Technical process: Specifies not only what needs to be done, and in what order, but also provides specific details of *how* to execute the various tasks.

Technically focused brainstorming: The use of standard brainstorming methods bounded by certain acceptable solution concept conditions and guided by the attainment of an *ideal solution*.

Technically focused brainstorming: This methodology guides the generation of solution concepts by ensuring that those solution concepts support the resolution of contradictory requirements of the system under analysis and renders that system to be of higher value than it was before the solution was applied.

Technology mapping and recombination: A matrix-based method that lists the various technologies that can perform a function and then examines combinations that have not been tried to see if there is enhanced performance or features.

Theory development and conceptual thinking: These people love thinking and talking about abstract ideas. They love the *why* of strategy more than the *how*. They may enjoy business models that explain the reasons behind the competitive position of a business.

Things are not innovation: The following maintenance or change management activities are *not* types of innovation:
- Cost savings
- Ingredient or product changes
- Regulatory change
- Label change

Thinking innovatively: Thinking innovatively is soliciting ideas from everyone, which is a challenge. There is a need for training people in asking questions, thinking of ideas, and articulating their ideas in words or graphics.

Thrashing: A term used to describe ineffective human workgroup activity, effort lost in unproductive work.

Time-of-day map: This tool focuses the participants not on a task or an experience, but rather what happens or does not happen in two- to four-hour chunks in a person's day and what opportunities may appear.

Top–down planning for innovation: Generally a revenue goal-driven process that is usually set from the top by the senior leadership team. It is usually a dollar revenue goal or a percentage of revenue target from innovation.

Trademarks: Words or logos that are used by someone to identify their products or services and distinguish them from the words or logos of others.

Trade secrets: Essentially refer to the legal protection often granted to confidential information having at least potential competitive value.

Tree diagram breakdown (drilldown): *A technique for breaking complex opportunities and problems down into progressively smaller parts. Start by writing the opportunity statement or problem under investigation down on the left-hand side of a large sheet of paper. Next, write down the points that make up the next level of detail a little to the right of this. These may be factors contributing to the issue or opportunity, information relating to it, or questions raised by it. For each of these points, repeat the process. This process of breaking the issue under investigation into its component part is called drilling down.*

Trends: Those dimensions on which lead users are far ahead of the mass market.

Trial and error: Attempts at successful solutions to a problem with little benefit from failed attempts.

Value analysis: *The analysis of a system, process, or activity to determine which parts of it are classified as real-value-added (RVA), business-value-added (BVA), or no-value-added (NVA).*

Value proposition: *A value proposition is a document that defines the benefits that will result from the implementation of a change or the use of an output as viewed by one or more of the organization's stakeholders. A value proposition can apply to an entire organization, parts thereof, or customers, or products, or services, or internal processes.*

Venture capitalist: Secures funding for innovation, evaluates and selects projects to receive resources, and guides implementation.

Vision: *A documented or mental description or picture of a desired future state of an organization, product, process, team, a key business driver, activity, or individual.*

Vision statements: *A group of words that paints a clear picture of the desired business environment at a specific time in the future. Short-term vision statements usually are between three and five years. Long-term vision statements usually are between 10 and 25 years. A visions statement should not exceed four sentences.*

Zones of conflict: Refers to the temporal zone and the operating zone of the problem—loosely the time and space in which the problem occurs.

Index

Page numbers followed by f, t and n denote figures, tables and footnotes, respectively.